CISM COURSES AND LECTURES

The series presents lecture notes, monographs, edited works and proceedings in the field of Mechanics, Engineering, Computer Science and Applied Mathematics.
Purpose of the series is to make known in the international scientific and technical community results obtained in some of the activities organized by CISM, the International Centre for Mechanical Sciences.

INTERNATIONAL CENTRE FOR MECHANICAL SCIENCES

COURSES AND LECTURES - No. 370

FLOW OF PARTICLES IN SUSPENSIONS

EDITED BY

U. SCHAFLINGER
TECHNICAL UNIVERSITY OF VIENNA

Springer-Verlag Wien GmbH

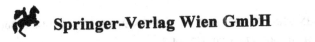

Le spese di stampa di questo volume sono in parte coperte da
contributi del Consiglio Nazionale delle Ricerche.

This volume contains 65 illustrations

Originally published by Springer-Verlag Wien New York in 1996

In order to make this volume available as economically and as
rapidly as possible the authors' typescripts have been
reproduced in their original forms. This method unfortunately
has its typographical limitations but it is hoped that they in no
way distract the reader.

ISBN 978-3-211-82813-7 ISBN 978-3-7091-2714-8 (eBook)
DOI 10.1007/978-3-7091-2714-8

PREFACE

The flow of solid particles, liquid droplets or gaseous bubbles suspended in a continuous liquid phase are typical multi-phase problems and cover a wide range of interesting applications, such as: cellulose fibers suspended in water; glass fibers in liquid polymers; deformable capsules; bubbles and droplets suspended in a flowing liquid; colloidal suspensions; sedimentation and transport of particles as well as fluidization phenomena.

This volume presents the lectures given during the 1995 course on the "Flow of particles in suspensions" held at the International Centre for Mechanical Sciences in Udine. Topics include the structure, the macroscopic transport properties and thermodynamic aspects of suspensions, the motion of capsules in shear flows, the effect of surfactants on the motion of bubbles and drops, the laminar transport of solid particles suspended in liquids and the separation of non-colloidal suspension under action of gravity or centrifugal forces. The objective of the course was to offer professionals and graduate students an up-to-date overview of issues related to the flow of particles in suspensions. The lectures not only outlined advanced theoretical approaches but also recent experimental techniques.

The first part of the volume focuses on physical comprehension and macroscopic modeling of multi-phase mixtures. The macroscopic transport properties depend on the micro-structure which is in shear-induced non-equilibrium steady states markedly different from equilibrium. Also, a macroscopic description of suspensions that can be applied to colloidal dispersions and molecular mixtures as well is presented.

The second part of the volume seeks to give insight into the mechanics of deformation of capsules and cells with either an elastomer or bilayer membrane suspended in a flowing liquid. It also aims to explain interfacial phenomena such as the adsorption of surface active agents (surfactants) and their influence on the motion of gas bubbles and liquid droplets through a continuous fluid.

In the last part of the volume a hydrodynamic diffusion model is presented to analyze the overall transport of solid particles in a laminar shear flow. Also a theory based on averaging techniques is outlined to predict separation of the two phases caused by gravity or centrifugal forces. In connexion with the classic theory of rotating fluids some new challenging questions are discussed.

We would like to especially thank Prof. Cognet for his help in co-organizing the program.

The authors are grateful to all members of CISM and in particular to Prof. Bianchi, Prof. Kaliszky, Prof. Schneider and Prof. Tasso for their advice and support in preparing both the 1995 course and this volume.

U. Schaflinger

CONTENTS

Page

STRUCTURE AND TRANSPORT PROPERTIES OF COLLOIDAL SUSPENSIONS IN STATIONARY SHEAR FLOW

J. Blawzdziewicz
Polish Academy of Sciences, Warsaw, Poland

ABSTRACT

The microstructure in colloidal suspensions can be strongly distorted by a moderate shear flow. Macroscopic transport properties depend on the microstructure and, therefore, are markedly different in equilibrium and in shear-induced nonequilibrium steady states. Experimental and theoretical results are reviewed for colloidal suspensions in stationary shear flow. The shear-rate-dependent stress, mobility, and diffusion tensors and the distortion of the static structure factor are discussed. A detailed microscopic analysis of a semidilute suspension is presented.

1. INTRODUCTION

The main objective of these lectures is to review important phenomena occurring in colloidal suspensions in stationary shear flow.

Structural relaxation times in colloidal suspensions are about nine orders of magnitude longer than in molecular fluids. Because of this slow relaxation, even a moderate shear flow may strongly influence suspension microstructure producing states far from thermodynamic equilibrium. Suspension transport properties are related to the microstructure and therefore depend strongly on the flow.

We consider suspensions of spherically symmetric rigid particles. Thus, the correlations of particle positions are sufficient to describe the flow-induced structural changes. The correlations are determined by the competition of Brownian motion and shear-induced particle convection.

Shear-induced nonequilibrium steady states are anisotropic even in suspensions with isotropic equilibrium states. Consequently, the static structure factor (measured in light and neutron scattering experiments) is not spherically symmetric, normal stress differences are nonzero, and effective particle mobilities and diffusivities are tensors rather than scalars.

We give a broad review of experimental and theoretical results and a detailed analysis of a simple theoretical model for a semidilute suspension of charged colloidal particles in a stationary shear flow. In Chapter 2, fundamental time-scales and basic properties of colloidal suspensions are reviewed. Chapter 3 examines the changes of structure and stress tensor in nonequilibrium shear-induced steady states. Chapter 4 focuses on particle migration resulting from perturbations of a steady state caused by external forces or concentration gradients. In Chapter 5, the results are discussed and directions for further research are suggested.

2. COLLOIDAL SUSPENSIONS

2.1 Fundamental Time Scales

The steady-state nonequilibrium structure of a colloidal suspension in shear flow is a function of two competing processes. The first is Brownian motion, responsible for relaxation of the system towards equilibrium. The second is particle migration induced by the flow that drives the system out of equilibrium. An estimate for the magnitude of the shear-induced structure distortion is obtained by comparing characteristic time scales associated with these two processes.

Brownian motion is characterized by the diffusion constant of an isolated particle given by the Einstein relation $D_0 = k_B T \mu_0$, where k_B is the Boltzmann constant, T is temperature, and μ_0 is the isolated-particle mobility. The characteristic time τ_B of Brownian structural relaxation is equal to the time for an isolated particle to diffuse

actual radius	[nm]	1	10	100	1000
τ_V	[s]	2×10^{-13}	2×10^{-11}	2×10^{-9}	2×10^{-7}
τ_B	[s]	1×10^{-7}	1×10^{-4}	1×10^{-1}	1×10^{2}
τ_{St}	[s]	2×10^{3}	2×10^{2}	2×10^{1}	2×10^{0}
τ_V/τ_B		2×10^{-6}	2×10^{-7}	2×10^{-8}	2×10^{-9}
τ_B/τ_{St}		5×10^{-11}	5×10^{-7}	5×10^{-3}	5×10^{1}

Table 1, Characteristic time scales for colloidal particles with different size. Effective particle radius is 5 times larger than actual radius.

a distance equal to its effective radius b based on interactions with other particles:

$$\tau_B = \frac{b^2}{D_0}. \tag{1}$$

The time scale τ_{sh}, associated with the structural changes induced by the flow, is the inverse of the shear rate:

$$\tau_{\text{sh}} = \gamma^{-1}. \tag{2}$$

The relative influence of shear flow on the suspension structure can be characterized by the Peclet number:

$$Pe = \frac{\tau_B}{\tau_{\text{sh}}}. \tag{3}$$

A separate Peclet number is associated with external forces acting on colloidal particles. Structural changes induced by an external force F^{ext} are characterized by the Stokes time τ_{St} which is the time for an isolated particle to travel a distance b:

$$\tau_{\text{St}} = \frac{b}{\mu_0 F^{\text{ext}}}. \tag{4}$$

The corresponding Peclet number Pe_{ext} is defined as

$$Pe_{\text{ext}} = \frac{\tau_B}{\tau_{\text{St}}}. \tag{5}$$

In these lectures, we consider situations where $Pe = O(1)$ or $Pe \gg 1$ and $Pe_{\text{ext}} \ll 1$. For nonuniform systems, we also assume that the particle concentration relaxes towards a uniform state on a time scale much larger than τ_B, which implies that spatial nonuniformities have a length scale much larger than b.

An estimate of the Brownian relaxation time τ_B and the Stokes time τ_{St} for aqueous suspension of spherical particles of different sizes is given in Table 1. This estimate indicates that intermediate and large Pe can be achieved at moderate shear rates (e.g., $\gamma = 100\,\mathrm{s}^{-1}$ for colloidal particles of actual radius $\gtrsim 0.1\,\mu$m.) This is in contrast to molecular fluids for which shear rates larger by 9 orders of magnitude are needed to achieve $Pe = O(1)$. (Lower shear rates, however, are sufficient to distort the structure of fluid mixtures near critical point because concentration fluctuations are slowed down [1].)

We conclude this section by remarking that the structural relaxation in colloidal suspensions takes place on a much slower scale than the relaxation of the fluctuating velocity of individual Brownian particles. An estimate of such velocity relaxation time τ_V for particles of several sizes is given in Table 1. Due to a very short relaxation time only the evolution of spatial structure is observed in most experiments, not the fluctuating particle velocities. Moreover, as can be estimated from the time scales, the fluctuating velocity distribution is unaffected by shear. It is appropriate, therefore, to base theoretical descriptions on the Smoluchowski equation where the evolution in a system of interacting Brownian particles is described as a diffusive process.

2.2 Interactions of Colloidal Particles

Mutual interactions of colloidal particles are mediated by molecules present in the suspending fluid and thus, depend strongly on its composition (e.g., ionic strength, presence of polymer chains). The relaxation time for non-hydrodynamic modes of the fluid is much shorter than the Brownian relaxation time τ_B; therefore, the influence of the suspending fluid can usually be described in terms of effective interactions between suspended colloidal particles (see, e.g., Ref. [2] and references therein). Effective interactions between colloidal particles include effective interparticle potentials and hydrodynamic interactions.

2.2.1 Interparticle Potentials

In general, the potential of interaction between more than two colloidal particles is not pair additive. However, pair additivity is often a reasonable approximation and is usually assumed for simplicity. Several examples of interparticle potentials follow (for more information see [3]–[5]):

Hard-core repulsion: This repulsion results from the rigid-body nature of solid colloidal particles.

Van der Waals attraction: This force results from mutual interactions of fluctuating electrical polarizations of the particles. It diverges upon contact and thus, colloidal particles tend to coagulate if no stabilizing repulsive force prevents their close approach.

Steric stabilization. Coating of colloidal particles with a layer of polymer results in strong, short-range particle repulsion. Depending on the range of the repulsion the interaction of the particles may be approximated by a hard-sphere potential or a hard

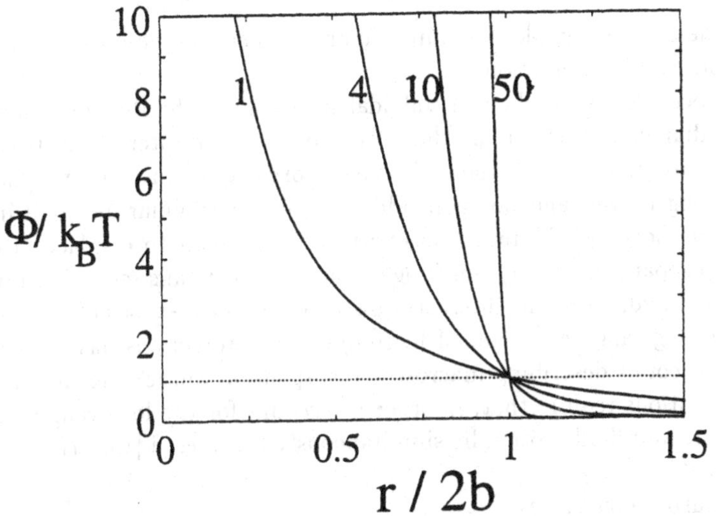

Fig. 1, Normalized Yukawa potential (6) plotted as function of normalized distance $r/2b$. Curves are labeled by value of hardness parameter α.

sphere with a deep narrow potential well of the van der Waals attraction.

Electrostatic repulsion. Coulombic interactions between charged colloidal particles are screened by ions in the solvent. The interactions are often well represented by the Yukawa potential,

$$\frac{\Phi(r)}{k_B T} = \frac{2b}{r} \exp\left[-\alpha\left(\frac{r}{2b} - 1\right)\right], \tag{6}$$

where r is the interparticle distance and $2b$ is the distance at which $\Phi = k_B T$. The parameter α describes "hardness" of the potential. In the limit $\alpha \to \infty$, $\Phi(r)$ tends to a hard sphere potential with radius b. For strongly charged colloidal particles suspended in a weakly ionic solvent, the effective radius b may greatly exceed the actual radius a. For such systems, b is the proper spatial scale for the suspension microstructure and is used to define time-scales (1) and (4). Hydrodynamic interactions play a relatively minor role for $b \gg a$. Yukawa potential for several values of parameter α is plotted in Fig. 1.

2.2.2 Hydrodynamic Interactions

Particles immersed in a viscous fluid do not move independently. The fluid displaced by a particle interacts with other particles, resulting in complex, many-body hydrodynamic interactions. Herein, for the sake of simplicity, we consider a suspension model where hydrodynamic interactions are negligible because of strong long-range electrostatic repulsion which prevents the close approach of particles. However, in order to provide a more complete picture, we briefly mention the most important features of hydrodynamic interactions.

For typical processes taking place in colloidal suspensions, the Reynolds number based on particle diameter is very small; thus, the hydrodynamic interactions between colloidal particles are described by the stationary Stokes equations. Hydrodynamic interactions are long ranged and not pair additive and are singular at small interparticle separations because of lubrication effects (it is difficult to displace viscous fluid from the gap separating nearly touching surfaces). For details regarding fundamental properties of hydrodynamic interactions see [6] and for methods of evaluation see [7]–[15]. The long range character of hydrodynamic interactions has important consequences for theories describing macroscopic suspension motion: local effective macroscopic descriptions require that, in order to account for the long-range terms, the average particle and fluid motion be simultaneously considered [16, 17].

2.3 Equilibrium Suspension States

Equilibrium suspension states are fully determined by the effective interparticle potentials and can be described using standard methods of statistical physics. In many respects, equilibrium properties of colloidal suspensions are similar to those of the molecular systems (for review see [5, 18]).

Colloidal suspensions, like simple fluids, may exhibit phase transitions from a fluid-like disordered phase to an ordered, crystalline solid-like state. For hard spheres, the transition occurs at a volume fraction of about 0.49. The crystallization volume fraction is a good indication of the range of the repulsive potential between colloidal particles. For strongly charged colloidal particles suspended in deionized water, the phase transition may occur at a volume fraction of solids below 0.1% indicating that the ratio of the effective to actual particle radius is $O(10)$.

3. NONEQUILIBRIUM STRUCTURE AND STRESS TENSOR IN SHEARED COLLOIDAL SUSPENSIONS

3.1 Introduction

As explained in Sec. 2.1, nonequilibrium states corresponding to intermediate and large Peclet numbers can easily be achieved in sheared colloidal suspensions. From the estimate of time scales, it follows that the flow may induce large distortions of the suspension microstructure for $Pe \gtrsim O(1)$. Such distortions are readily observed in experiments. Since macroscopic transport in colloids depends on microstructure, Peclet-number dependence of suspension transport properties is expected.

In this chapter, we consider properties of the nonequilibrium steady state for a uniform sheared suspension. We describe the flow-induced structural changes observed in light scattering experiments and characterize the Peclet-number dependence of the

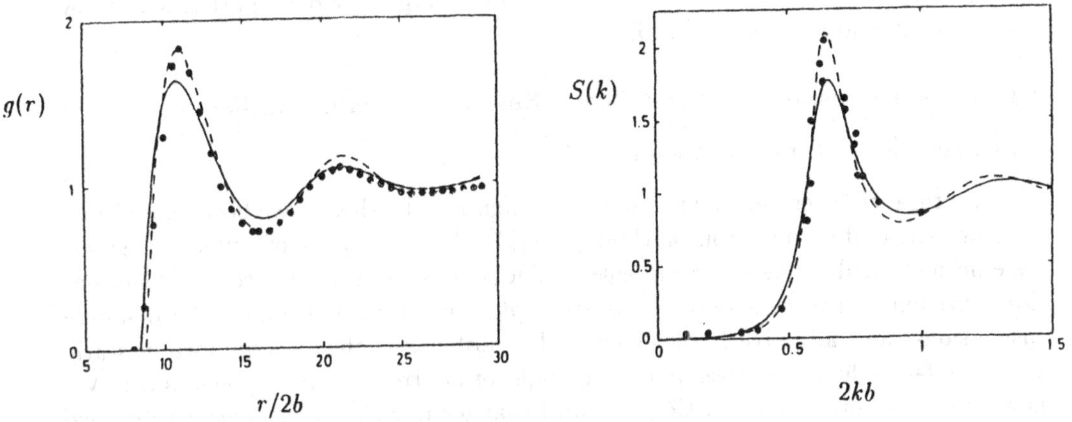

Fig. 2, Equilibrium pair distribution function and static structure factor for charged colloidal particles of actual radius $a = 23\,\text{nm}$, surface potential 150 mV, and volume fraction 4.4×10^{-4}, at low electrolyte concentration. Data points: computer simulations; lines: integral-equation results. Note that effective radius $\sim 10a$. From Nägele et al. [19]; system parameters from Pusey [5].

suspension effective viscosity and other macroscopic rheological coefficients. In Sec. 3.2, we define the static structure factor—a standard quantity for characterizing the suspension structure. Next, in Sec. 3.3 we review interesting experimental findings. A theoretical discussion of the observed phenomena is given in Sec. 3.4.

3.2 Characterization of Suspension Structure

An important function characterizing the structure of colloidal suspensions is the two-particle density $n_2(\mathbf{r}_1, \mathbf{r}_2)$: the density of pairs of colloidal particles with the centers at points $\mathbf{r}_1$ and $\mathbf{r}_2$. For statistically uniform suspensions, the pair distribution function $g(\mathbf{r})$ is defined by the relation $g(\mathbf{r}) = n_2(\mathbf{r}_1, \mathbf{r}_2)/n^2$, where n is the mean particle number density and $\mathbf{r} = \mathbf{r}_1 - \mathbf{r}_2$. In fluid-like isotropic equilibrium states, the pair distribution g depends only on the relative distance of the particles $r = |\mathbf{r}|$. In contrast, nonequilibrium states of a suspension in shear flow are anisotropic so that $g_2(\mathbf{r})$ has a nontrivial angular dependence.

A closely related quantity, the static structure factor

$$S(\mathbf{k}) = 1 + n \int d^3r\, [g(\mathbf{r}) - 1] \exp(i\mathbf{k}\cdot\mathbf{r}), \tag{7}$$

is measured in light, X-ray, and neutron scattering experiments [5, 18]. Like the pair density, $S(\mathbf{k})$ in fluid-like equilibrium states is isotropic but out of equilibrium it usually depends on the direction of $\mathbf{k}$. The equilibrium pair distribution function and the static structure factor for a suspension of charged colloidal particles at a low ionic concentration are plotted in Fig. 2.

3.3 Sheared Suspension in a Steady State: Review of Experimental Results

3.3.1 Distortion of Structure Factor

Structural changes induced in colloidal suspensions by shear have been experimentally investigated by numerous authors [20]–[31]. A wide variety of colloidal systems have been studied in these experiments, including charge stabilized colloidal suspensions with long-range screened electrostatic repulsion, sterically stabilized hard-sphere suspensions, and adhesive hard spheres. In most cases, the nonequilibrium static structure factor has been observed by the light or neutron scattering techniques. We now describe some characteristic structural changes in sheared suspensions revealed by these experiments.

Shear induced distortion of the liquid-like colloidal structure was observed in a pioneering work by Clark and Ackerson [21] using light scattering. Later, a more detailed study of a similar system was reported by Ackerson [23]. A typical system used in these experiments was a dilute deionized aqueous suspension of $0.109\,\mu$m diameter polystyrene spheres. The concentration of solids was only about 0.1% but due to strong electrostatic repulsion the suspension exhibited a pronounced equilibrium structure characteristic of a dense liquid-like state. The estimate of the effective diameter of the particles, based on the position of the first maximum in the equilibrium static structure factor, was of the order of $2b = 1\,\mu$m.

A qualitative picture of the structural changes observed by Clark and Ackerson is given in Fig. 3. The equilibrium static structure factor exhibits a characteristic circular Debye-Scherrer ring (Fig. 3a). At nonzero shear rates, the scattering pattern is stretched along the compression axis and compressed along the extensional axis (Fig. 3b). This change corresponds to the compression of the pair correlation function along the compression axis and stretching along the extensional axis.

The observed intensity of scattered light along the distorted Debye-Scherrer ring was nonuniform. The magnitude of the ring distortion and the magnitude of the scattered light nonuniformity along the ring increased with the shear rate. This effect is illustrated quantitatively in Fig. 4. The Peclet number used in the analysis of the data was based on the effective particle diameter $2b$.

Analogous results have also been obtained for the static structure factor measured in different directions with respect to the flow. For example, Yan and Dhont [30] reported measurements in the velocity-vorticity plane for slightly charged silica particles suspended in a mixture of ethanol and toluene. The crystallization volume fractions for this system were $\phi = 12.6\%$ (fluid phase) and 13.2% (crystalline phase). They found a large distortion in the flow direction and a small distortion in the vorticity direction. The latter is significant because the widely used mode-mode coupling approximation predicts a zero distortion in the plane perpendicular to the flow.

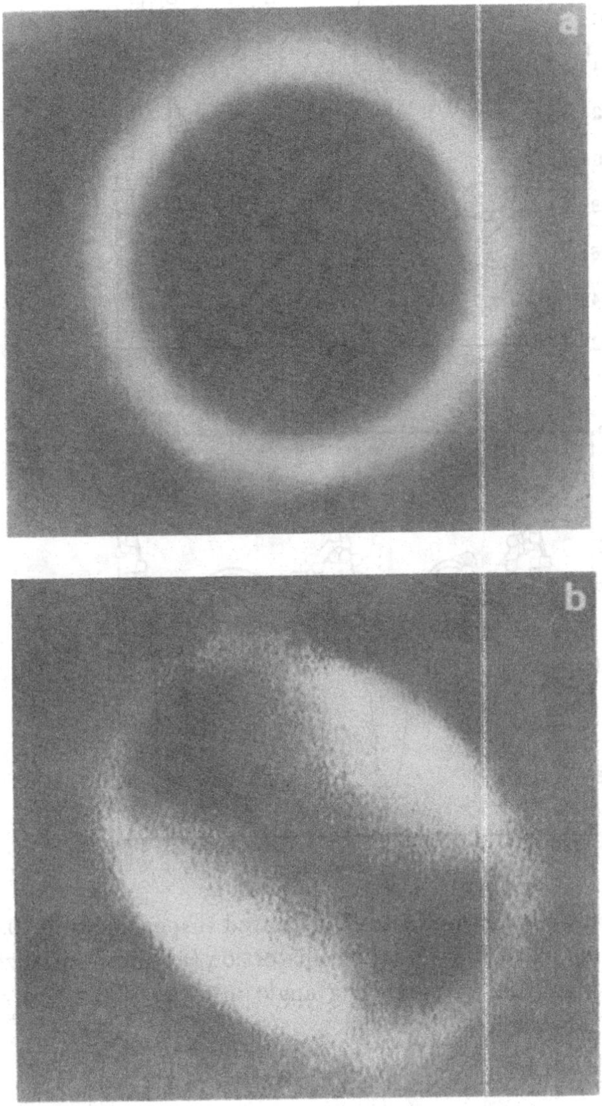

Fig. 3, Shear induced distortion of static structure factor in shear plane for colloidal hard sphere suspension with volume fraction 0.3. (a) Equilibrium structure factor (predicted by Percus-Yevick theory). (b) Ratio of sheared to equilibrium structure factor for polymethylmethacrylate "hard" spheres. Shear direction is horizontal, velocity direction is vertical. From Ackerson [23].

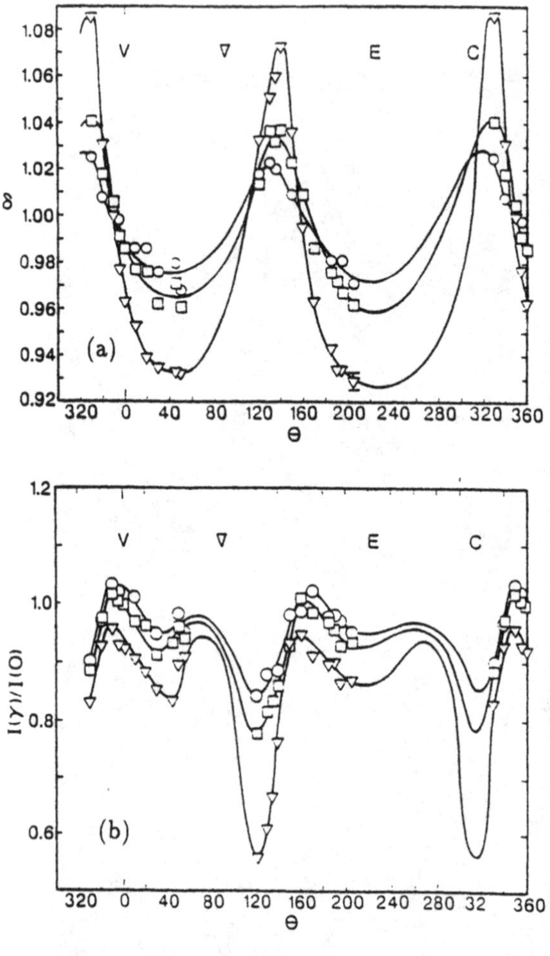

Fig. 4, Main peak of static structure factor for sheared suspension of $0.109\,\mu$m diameter charged colloidal particles. Relative shear distortion (a) and normalized scattered light intensity (b) as functions of scattering angle in shear plane; Pe = 0.37 (○), 0.69 (□), and 3.2 (▽). From Ackerson [23].

The experiments discussed above were performed at particle concentrations for which equilibrium suspension states were fluid-like. The structure factor was distorted by shear but long-range order was not observed. In suspensions that have an ordered crystalline equilibrium state, a long-range order has also been observed in steady shear flow [22], [24], [30]–[34]. The structure of such a sheared suspension ranges from a layered ordering at low Peclet numbers, to shear-induced partial translational string-

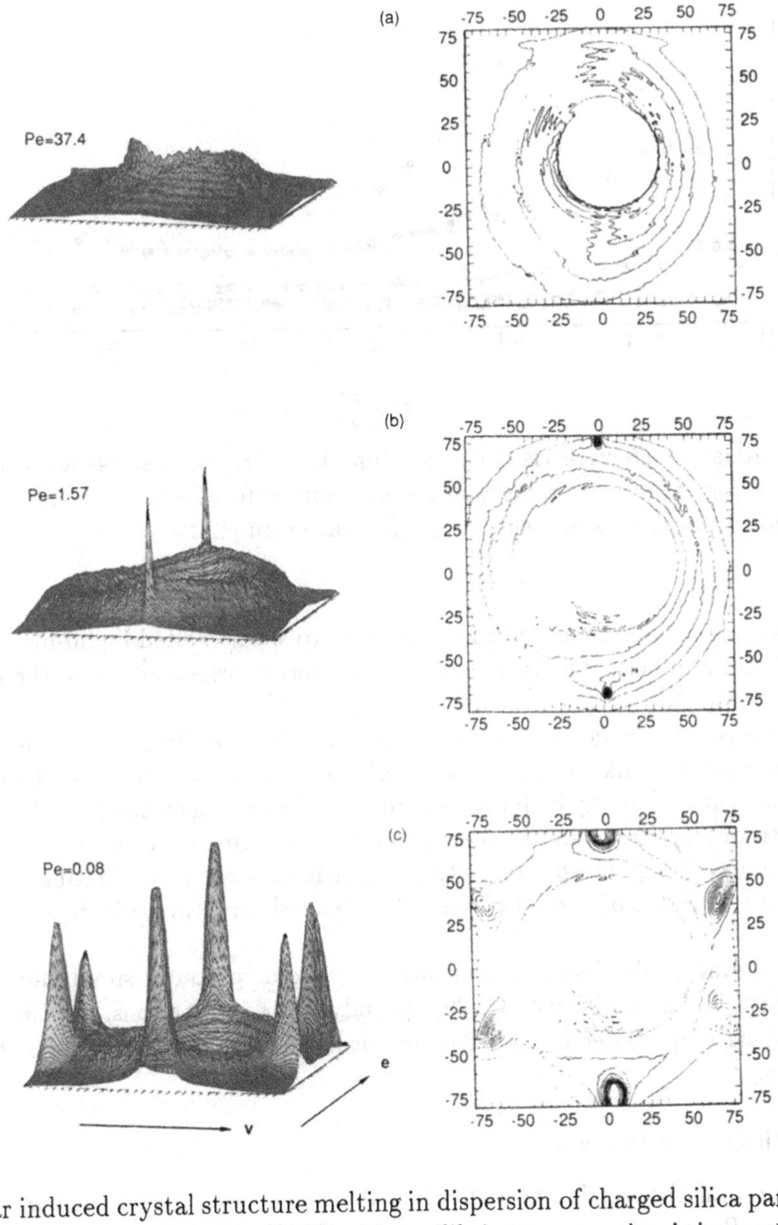

Fig. 5, Shear induced crystal structure melting in dispersion of charged silica particles with volume fraction of solids $\phi = 13.9\%$. At equilibrium, suspension is in crystalline state for $\phi \geq 13.2\%$. Light scattering patterns are in velocity-vorticity plane. (a) Shear-melted liquid-like order. (b) Shear-induced string-like ordering. (c) Shear-induced layering. From Yan and Dhont [30].

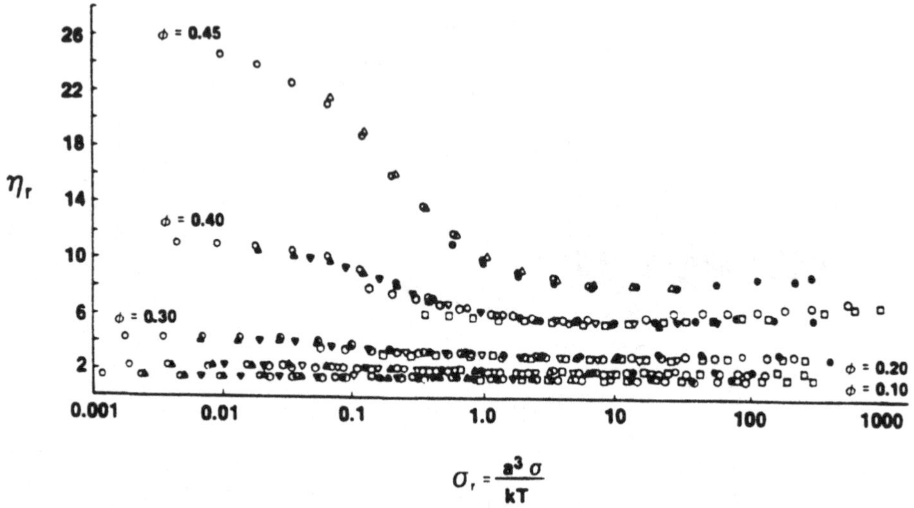

Fig. 6, Relative effective viscosity as a function of reduced stress for suspensions of polymethylmethacrylate spheres at several volume fractions. Different symbols refer to various sphere sizes and suspension media. From [42].

like ordering at intermediate Peclet numbers, to shear-melted liquid-like structure at high Peclet numbers. The light-scattering patterns corresponding to these states are shown in Fig. 5.

There is some evidence that for suspensions that are fluid-like in the equilibrium state, a crystalline-like interparticle ordering can be induced by oscillatory shear flow [22, 35, 36] at sufficiently high Peclet numbers. Depending on the particle volume fraction, strain amplitude, and frequency, several scattering patterns are possible. Some patterns have been interpreted in terms of twin-crystalline, oscillating fcc structures and some in terms of a randomly-stacked layered structure oriented along the flow axis.

Long range order has also been predicted from computer simulations of suspensions with [37] and without [38, 39] hydrodynamic interactions. Further discussion of the long-range order in colloidal suspensions is beyond the scope of the present lectures.

3.3.2 Rheological Behavior

Structural changes in sheared colloidal suspensions are accompanied by a characteristic Peclet number dependence of the effective viscosity. Measurements of the effective viscosity have been reported for many different colloidal systems and for a wide range of Peclet numbers [20],[31], [40]–[44]. Useful reviews of this topic can be found in Refs. [3, 5] and the references therein. We briefly recall some characteristic results.

Let σ_{ij} be the macroscopic stress tensor in a suspension undergoing a stationary shear flow of the form $\mathbf{v} = \gamma y \hat{\mathbf{e}}_x$. The effective viscosity η^{eff} is defined by the relation:

$$\sigma_{xy} = \eta^{\text{eff}} \gamma. \tag{8}$$

Choi and Krieger [42] have measured the effective viscosity of a suspension of polymethacrylate spheres stabilized by coating with a thin polymer layer. Their results for the relative effective viscosity $\eta_r = \eta^{\text{eff}}/\eta$ (where η is the viscosity of the suspending fluid) are plotted in Fig. 6 for different volume fractions as a function of the reduced stress $s_r^* = a^3\sigma_{xy}/k_B T$ (which is roughly proportional to the Peclet number (3) [5]).

According to Fig. 6, the effective viscosity is nearly constant for small volume fractions $\phi \leq 0.2$. At higher volume fractions, it decreases substantially with shear-rate beyond a small Peclet-number Newtonian plateau. This phenomenon is known as shear thinning. At large Peclet numbers the viscosity is again approximately constant. This regime is often incorrectly called the "second Newtonian plateau." In fact, in this regime the suspension is not Newtonian, with nonzero normal stress differences $N_1 = \sigma_{xx} - \sigma_{yy}$ and $N_2 = \sigma_{yy} - \sigma_{zz}$, and the shear viscosity depending on the form of the flow. At the highest volume fractions considered by Choi and Krieger, one observes a slight increase of the effective viscosity at very high Peclet numbers (shear thickening).

Rheology of suspensions undergoing shear-induced long-range-ordering transitions is much more complex (see [4, 31, 45] and references therein). Measurements of normal stresses in suspensions of colloidal particles in Newtonian fluids are scarce [46]–[48].

3.4 Sheared Suspension in a Steady State: Some Theoretical Considerations

In the preceding section, we have reviewed experimental observations of nonequilibrium steady states in sheared colloids. A complete theoretical description of such states is still lacking; in recent years, however, substantial theoretical progress has been made. Earlier results include the theories of Ronis [49], Schwarzl and Hess [50], and Dhont [51]. Bławzdziewicz and Szamel [52] explain that these approximate theories yield only qualitative results. The argument is based on an accurate solution of the Smoluchowski equation for a simple suspension model, in the small-to-intermediate Peclet-number regime. This analysis is described in subsections 3.4.1–3.4.4 and supplemented in subsection 3.4.5 by a discussion of the large Peclet number limit. The results of Bławzdziewicz and Szamel have recently been extended in several ways [53, 54] (see discussion in Chapter 5).

3.4.1 Simple Suspension Model

We now introduce the suspension model for which shear-induced structure was analyzed in [52]. A similar model is also used in the analysis of particle migration in sheared suspensions [55, 56], presented in Chapter 4.

The suspension consists of identical, spherically-symmetric Brownian particles with the effective radius b interacting via a pair-additive potential $\Phi(r)$. In the calculations presented in this section a hard-sphere potential is assumed. However, in Chapter 4 some related results will also be given for continuous $\Phi(r)$. It is assumed that b is much larger than the actual particle radius and therefore, the hydrodynamic interactions between the particles are neglected. This is a reasonable approximation for strongly charged colloidal suspensions in weakly ionic solvents (cf. Subsec. 2.2.1–2.2.2). The analysis is for low densities, so only two-particle contributions to the structure factor and transport coefficients are included. The system is statistically uniform.

The suspension undergoes a stationary shear flow of the form:

$$\mathbf{v}(\mathbf{r}) = \gamma y \hat{\mathbf{e}}_x. \tag{9}$$

In the absence of hydrodynamic interactions, a particle at a point $\mathbf{r}_1$ acquires a drift velocity $\mathbf{v}(\mathbf{r}_1)$ and undergoes Brownian motion with the diffusion constant of an isolated particle D_0. In the low density limit, the shear-induced distortion of the pair distribution $g^{st}(\mathbf{r})$ can be determined from the stationary two-particle Smoluchowski equation [3]. For our model, the Smoluchowski equation written in the relative-position coordinates $\mathbf{r}$ is:

$$-\nabla \cdot [-2D_0\nabla + \gamma y \hat{\mathbf{e}}_x]\, g^{st}(\mathbf{r}) = 0 \text{ for } r > 2b, \tag{10}$$

with zero flux boundary condition for contact configurations:

$$\hat{\mathbf{r}} \cdot [-2D_0\nabla + \gamma y \hat{\mathbf{e}}_x]\, g^{st}(\mathbf{r}) = 0 \text{ at } r = 2b, \tag{11}$$

where $\hat{\mathbf{r}} = \mathbf{r}/r$.

3.4.2 Solution by Induced Sources Method

The Smoluchowski equation (10)–(11) has been solved [52] by an induced sources method. In this method, the solution of Eq. (10) in the whole space is represented as a linear combination of fundamental solutions corresponding to multipole sources inserted at $\mathbf{r} = 0$. The strength of the sources is adjusted to satisfy boundary condition (11) at $r = 2b$. We briefly recall some details of this procedure.

First a set of fundamental solutions $T_{\alpha\beta}$ corresponding to the multipole sources $(-\partial/\partial x)^\alpha(-\partial/\partial y)^\beta \delta^3(\mathbf{r})$ is constructed. These solutions can be expressed as:

$$T_{\alpha\beta}(\mathbf{r}) = (-1)^{\alpha+\beta} \int_0^\infty d\tau \left(\frac{\partial}{\partial x}\right)^\alpha \left(\frac{\partial}{\partial y} + 2\tau \frac{\partial}{\partial x}\right)^\beta P(\mathbf{r}, \tau) \tag{12}$$

where $P(\mathbf{r},\tau)$ is the Green function of the time dependent diffusion-convection problem [57]. Next, g^{st} is written as a linear combination of the fundamental solutions $T_{\alpha\beta}$:

$$g^{st}(\mathbf{r}) = 1 + \sum_{\alpha,\beta=0}^{\infty} C^{\alpha\beta} T_{\alpha\beta}(\mathbf{r}) \text{ for } r > 2b, \tag{13}$$

where the coefficients $C^{\alpha\beta}$ are determined by the boundary condition (11). By inserting (13) into (11), the boundary condition is transformed into the following form:

$$\sum_{\alpha,\beta=0} C^{\alpha\beta} J_{\alpha\beta}(\hat{\mathbf{r}}) = -\gamma \left.\frac{xy}{r}\right|_{r=2b}, \tag{14}$$

where

$$J_{\alpha\beta}(\mathbf{r}) = \hat{\mathbf{r}} \cdot [-2D_0\nabla + \gamma y \hat{\mathbf{e}}_x]\, T_{\alpha\beta}(\mathbf{r}) \tag{15}$$

is the radial particle current.

In order to find g^{st}, Eq. (14) must be solved for C. To this end, an expansion in Cartesian spherical harmonics $Y_{ab}(\hat{\mathbf{r}})$ is used:

$$J_{\alpha\beta}(\hat{\mathbf{r}}) = \sum_{a,b=0}^{\infty} J_{\alpha\beta}^{ab} Y_{ab}(\hat{\mathbf{r}}), \tag{16}$$

where

$$Y_{ab}(\hat{\mathbf{r}}) = (-1)^{a+b} r^{a+b+1} \frac{1}{[2(a+b)-1]!!} \frac{\partial^a}{\partial x^a} \frac{\partial^b}{\partial y^b} \frac{1}{r}. \tag{17}$$

Due to the symmetry of the problem it is sufficient to use a complete set of harmonics symmetric in z. Upon expansion, boundary condition (14) takes the form of an infinite set of algebraic equations for $C^{\alpha\beta}$:

$$\sum_{\alpha,\beta=0}^{\infty} J_{\alpha\beta}^{ab} C^{\alpha\beta} = -2\gamma b \delta_{a1}\delta_{b1}. \tag{18}$$

The set of equations (18) has been solved numerically, with the coefficients $J_{\alpha\beta}^{ab}$ calculated by two methods: (a) numerical integration and (b) expansion in $Pe^{1/2}$. At this point, we note that the dependence of the pair distribution function on Pe is non-analytic since a perturbation in Pe of the stationary Smoluchowski equation is singular: for an arbitrary value of Pe there is always a region $r/2b > Pe^{-1/2}$ where the effect of the convection is not small when compared with the diffusion. The analytical structure of the solution can be deduced by rescaling $\mathbf{R} = (\mathbf{r}/2b)Pe^{1/2}$. After rescaling, the Peclet number only appears in the boundary condition of equations (10)-(11) and the expansion of g^{st} in powers of $Pe^{1/2}$ can be obtained from an expansion in powers of R.

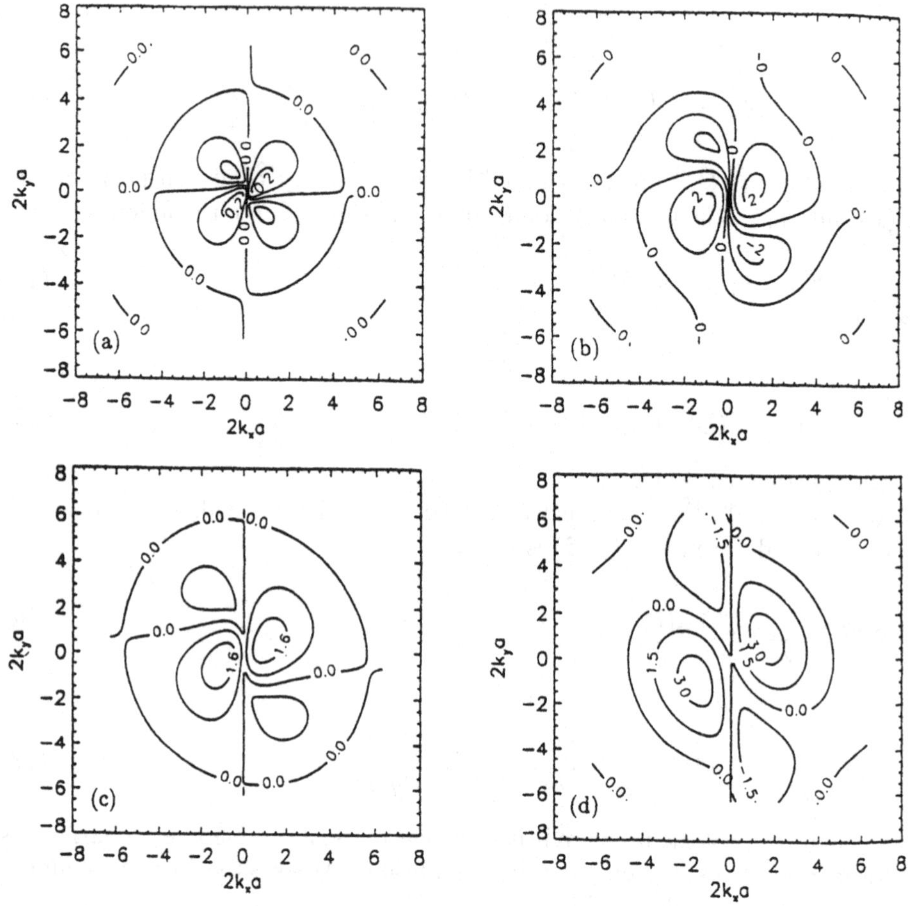

Fig. 7, Contour maps of $[S(\mathbf{k}, Pe) - S^{\mathrm{eq}}(k)]/\phi$ in gradient-velocity plane. (a) $Pe = 0.1$: present theory. $Pe = 2$: (b) present theory, (c) Dhont's approximation [51], (d) Schwarzl and Hess' theory [50]. From Bławzdziewicz et al. [52].

3.4.3 Structure Factor

Some results of the calculation described above are reproduced in Figs. 7–9. Since one of our goals was to assess the accuracy of earlier approximate theories [49]–[51], their predictions for the model system are also shown. Before discussing the details of our findings, we examine the approximations involved in the earlier theories [49]–[51]. Dhont's [51] approximate solution to the two-particle Smoluchowski equation

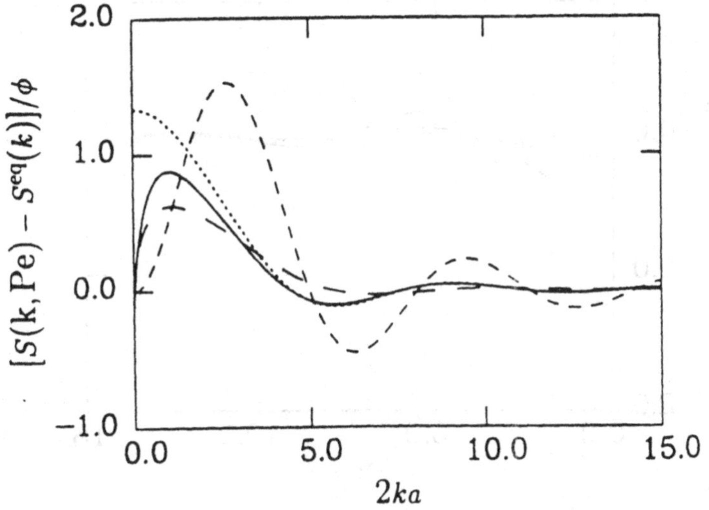

Fig. 8, Distortion $[S(\mathbf{k}, Pe) - S^{eq}(k)]/\phi$ of steady-state structure factor from equilibrium form, along dilatation axis $\mathbf{k} = k(\hat{\mathbf{e}}_x + \hat{\mathbf{e}}_y)/\sqrt{2}$, for reduced shear rate $Pe = 0.5$. Solid line: present theory; long-dashed: Dhont's approximation [51]; dashed: Schwarzl and Hess' approximation [50]; dotted: linear response result. From Bławzdziewicz et al. [52].

is equivalent to the mode-mode coupling approximation applied at low densities. It describes independent diffusion of two particles after an initial interaction, with no further interactions taken into account. The Ronis theory [49], which is based on the phenomenological fluctuation-diffusion equation rather than the Smoluchowski equation, is also equivalent to the mode-mode coupling approximation. In the low density limit the theories of Dhont and Ronis are equivalent. The simplification employed by Schwarzl and Hess [50] consists of replacing the diffusive term in the Smoluchowski equation with a term describing exponential relaxation with a single relaxation time.

The contour maps for the nonequilibrium distortion of the steady state structure factor $S(\mathbf{k})$ for $\mathbf{k}$ in the gradient-velocity plane are plotted in Fig. 7. The results are normalized by the effective volume fraction $\phi = \frac{4}{3}\pi b^3 n$. For $Pe = 0.1$, the distortion is relatively small and shows the symmetry of the x-y spherical harmonic. At a higher Peclet number, $Pe = 2$, the distortion is much larger and exhibits a more complicated angular dependence. For $Pe = 2$, the accurate solution of the Smoluchowski equation (10)–(11) differs considerably from the results predicted by the theories of Dhont and Schwarzl and Hess. We conclude that the approximate theories do not yield quantitative results, at least for the model system considered.

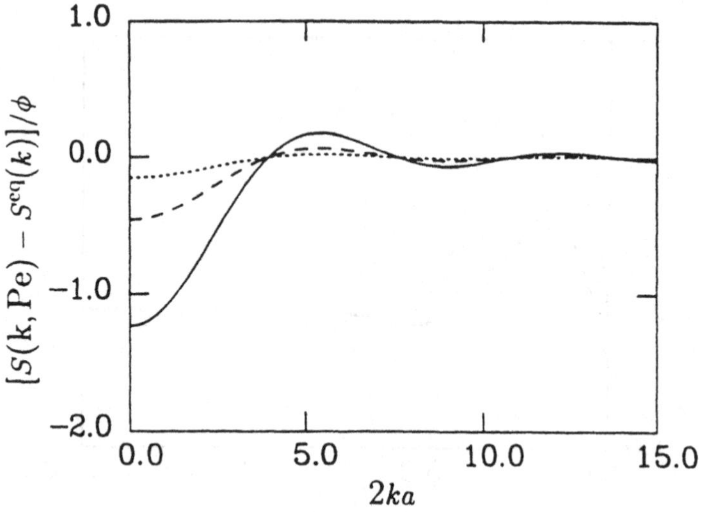

Fig. 9, Distortion $[S(\mathbf{k}, Pe) - S^{\mathrm{eq}}(k)]/\phi$ of steady-state structure factor from equilibrium form, in velocity gradient direction $\mathbf{k} = k\hat{\mathbf{e}}_y$. Dotted line: $Pe = 0.5$; dashed: $Pe = 1$; solid: $Pe = 2$.

A similar conclusion can be drawn from the results presented in Fig. 8, where the distortion of the structure factor along the extensional axis $\mathbf{k} = k(\hat{\mathbf{e}}_x + \hat{\mathbf{e}}_y)/\sqrt{2}$ is shown for $Pe = 0.5$. In this figure, the accurate solution to (10)–(11) is compared with the results based on approximations of [49]–[51]. The linear response result is also plotted. The linear response approximation correctly describes the large-k behavior of the distortion but fails for small k. This corresponds to the small-k boundary layer behavior of $S(\mathbf{k})$ discussed by Dhont [51].

In Fig. 9, distortion of the structure factor in the velocity gradient direction $\mathbf{k} = k\,\hat{\mathbf{e}}_y$ for several Pe is presented. In contrast with (10)–(11), the approximate theories [49]–[51] predict no distortion in directions perpendicular to the flow. However, a nonzero distortion has been experimentally observed [30].

The results presented in Figs. 7–9 cannot be directly compared to experimental observations since the theory is valid only for low volume fractions (based on b). Direct measurement of the structure factor at such low concentrations is very difficult. However, the theoretical predictions qualitatively agree with the continuous changes of the structure factor observed at higher densities.

3.4.4 Stress Tensor

Effective viscosity of suspensions is different from the viscosity of the suspending fluid because of the particle contribution to the stress. For a given concentration, the particle contribution depends on the suspension structure, which for rigid spherical particles is fully specified by correlations of their positions. The flow-induced distortion of the structure results in a nonlinear Peclet-number dependence of the stress tensor. Thus, the suspension behaves as a non-Newtonian fluid with a Peclet-number-dependent effective viscosity. A characteristic example was given in Subsec. 3.3.2. We now present a theoretical analysis of this effect for the simple suspension model introduced in Subsec. 3.4.1.

For a suspension with no hydrodynamic interactions, the particle contribution to the stress tensor consists of only two components: the first "perfect-gas" diagonal term $-nk_BT\mathbf{I}$ and the direct particle interaction term. For pair-additive interaction potentials, the two-particle distribution function $g^{\rm st}$ includes sufficient structural information to determine the stress. In particular, for hard-spheres, the direct-interaction contribution $\boldsymbol{\sigma}^{(2)}$ to the stress tensor can be written in the following form [3]:

$$\boldsymbol{\sigma}^{(2)} = -4n^2b^3k_{\rm B}T \int d^2\Omega\; \hat{\mathbf{r}}\,\hat{\mathbf{r}}\, g^{\rm st}(2b\hat{\mathbf{r}}). \tag{19}$$

For simple shear flow (9), the traceless part of the steady-state stress-tensor is fully characterized by three shear-rate dependent transport coefficients [58]:

$$\eta_+^{(2)} = \gamma^{-1}\,\sigma_{xy}^{(2)}, \tag{20}$$

$$\eta_-^{(2)} = \gamma^{-1}\,\frac{1}{2}\left(\sigma_{xx}^{(2)} - \sigma_{yy}^{(2)}\right), \tag{21}$$

$$\eta_0^{(2)} = \gamma^{-1}\,\frac{1}{2}\left[\sigma_{zz}^{(2)} - \frac{1}{2}\left(\sigma_{xx}^{(2)} + \sigma_{yy}^{(2)}\right)\right]. \tag{22}$$

In the low-shear-rate limit, $\eta_+^{(2)}$ is reduced to the two-body contribution to the Newtonian viscosity; and the coefficients of normal stress differences, $\eta_-^{(2)}$ and $\eta_0^{(2)}$, vanish.

The solution of the two-particle Smoluchowski equation (10)–(11), described in Subsec. 3.4.1, was used in [52] to calculate the rheological coefficients (20)–(22). In particular, an expansion in powers of $Pe^{1/2}$ was obtained and resummed with Pade approximants.

The dependence of the transport coefficients $\eta_+^{(2)}$, $\eta_-^{(2)}$, and $\eta_0^{(2)}$ on $Pe^{1/2}$ is shown in Fig. 10. For small Pe the coefficient $\eta_+^{(2)}$ is approximately constant but it decreases significantly with the increasing shear rate for Pe above approximately 0.3. At still higher shear rates, $\eta_+^{(2)}$ seems to approach a plateau, although the range of Peclet numbers considered is not sufficient to determine the limiting value at $Pe \to \infty$. The results of the calculations are in a qualitative agreement with the experimentally observed shear-thinning behavior of colloidal suspensions (cf. discussion in Subsec. 3.3.2). For $Pe \ll 1$, $\eta_-^{(2)}$ and $\eta_0^{(2)}$ vary as Pe. The coefficient $\eta_0^{(2)}$ increases monotonically in the range of Peclet numbers considered and $\eta_-^{(2)}$ has a maximum at $Pe = 4.2$.

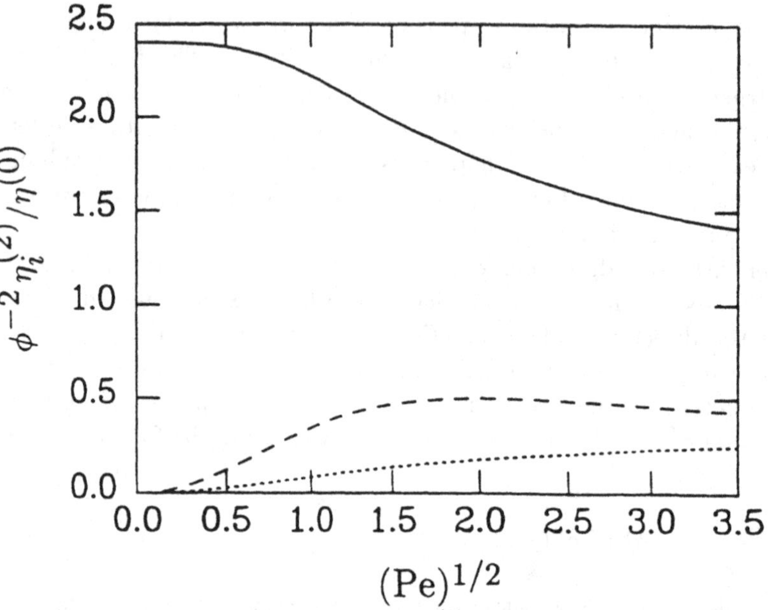

Fig. 10, Reduced viscosity coefficients $\eta_i^{(2)}$, defined in Eqs. (20)–(22), plotted as functions of $\sqrt{Pe}$. Solid line: $\phi^{-2}\,\eta_+^{(2)}/\eta_0$; dashed: $\phi^{-2}\,\eta_-^{(2)}/\eta_0$; dotted: $\phi^{-2}\,\eta_0^{(2)}/\eta_0$; where $\eta_0 = k_B T/(6\pi\, D_0\, b)$.

3.4.5 Large Peclet number limit

To supplement the results presented in Subsec. 3.4.1–3.4.4 we consider the case when there is no Brownian motion. In other respects the suspension model is the same as introduced in Subsec. 3.4.1 (semidilute hard spheres with no hydrodynamic interactions).

In such suspensions, an isolated particle follows the fluid streamlines until it comes into a contact with another particle. Pairs of particles in contact slide over each other until the center-to-center vector becomes perpendicular to the flow and the pair separates. The velocity of this sliding motion is determined by the condition that only center-to-center motion is constrained. The pair distribution in such a suspension can easily be calculated [4] under the assumptions that (i) binary particle encounters are uncorrelated and (ii) the particles never recollide. The pair distribution is equal to one everywhere except for a nonvanishing surface density of particles in contact and an infinite wake region corresponding to particles after separation.

The components of the stress tensor can be determined from particle contact forces, with the following result:

$$\eta_+^{(2)} = \frac{6}{5}\phi^2\eta^{(0)} \tag{23}$$

$$\eta_-^{(2)} = 0 \tag{24}$$

and

$$\eta_0^{(2)} = \frac{6}{5\pi}\phi^2\eta^{(0)}, \tag{25}$$

where $\phi = \frac{4}{3}\pi b^3 n$ is the volume fraction based on the effective radius b, and $\eta^{(0)} = (6\pi b\mu_0)^{-1}$.

The results (23)–(25) cannot be interpreted as the limit $Pe \to \infty$ of the Peclet-number dependent rheological coefficients (20)–(22) calculated from the Smoluchowski equation (10)–(11). Particle correlations in the wake region are destroyed at sufficiently large distances by arbitrarily weak Brownian diffusion. Collisions with other particles in the system would have a similar effect. Though less obvious, an arbitrarily weak Brownian motion for plane shear flow also causes a finite change of the pair distribution at *small* interparticle distances. The mechanism for this change is as follows:

For plane shear, particles drifting apart after their initial contact need to diffuse only a distance of the order of the particle radius b to enter streamlines bringing them back towards recollision. Since the return time and the time needed to enter a return streamline are of the same order, a finite fraction of the returning particles will meet again after the initial collision. A finite change of the pair distribution produces a corresponding finite change of the viscosity coefficients. In practice, a small perturbation of the plane shear by an extensional flow would remove the recollision mechanism at infinite Pe. However, the recollision effect may be significant for large but finite Peclet numbers.

Recently, Brady and Morris [59] have analyzed the structure of the pair distribution in the boundary layer for near-contact configurations, for large but finite Peclet numbers. To our knowledge, however, their solution does not include the recollision effect.

4. MOBILITY AND DIFFUSION IN SHEARED COLLOIDAL SUSPENSIONS

4.1 Introduction

This chapter focuses on particle migration processes produced by small perturbations around the nonequilibrium steady states of sheared colloidal suspensions, discussed in Chapter 3. Two kinds of perturbations are considered: those induced by a force acting on the suspended particles and those resulting from a gradient in the particle concentration. Particle transport in these two situations is characterized by the respective Peclet-number-dependent effective particle mobility and diffusion coefficients. As in the standard equilibrium case, we distinguish "collective" and "self" processes. The initial perturbation affects all particles in collective processes while it affects only the diluted labeled-particle component in self processes.

The discussion starts in Sec. 4.2 with macroscopic constitutive equations describing particle transport in sheared suspensions. In Sec. 4.3, we present the experimental study of Qiu et al. [60] where the influence of shear on the self-diffusion process in a colloidal suspension was investigated. Section 4.4 focuses on a detailed microscopic analysis of colloidal particle transport processes in shear flow using a simple suspension model similar to that described in Sec. 3.4.

4.2 Macroscopic Equations

Consider a steady state of a statistically uniform colloidal suspension in a stationary plane shear flow. Suppose this steady state is perturbed by a small external force $\mathbf{F}^{\text{ext}}$ and a small gradient of particle number density n_1. Collective and self perturbations are possible. For collective perturbations all particles are equivalent: the force $\mathbf{F}^{\text{ext}}$ acts on all of them, and n_1 refers to the total particle density. For self perturbations, some particles are tagged. The force $\mathbf{F}^{\text{ext}}$ acts only on the tagged particles, and the density of tagged particles $n_1 \equiv n_s$ is much smaller than the uniform density n of untagged particles. The tagged and untagged particles are otherwise identical.

On the macroscopic level, the time-evolution of the density n_1 in a sheared suspension can be expressed in terms of the continuity equation

$$\frac{\partial n_1}{\partial t} = -\nabla \cdot \mathbf{j}_1, \tag{26}$$

supplemented by the constitutive relation for the particle current $\mathbf{j}_1$ (total current in the collective case and the current of tagged particles in the self case). For small perturbations, long times, and $\mathbf{F}^{\text{ext}}$ and n_1 that vary slowly on the particle length scale, the constitutive relation is linear with the following form:

$$\mathbf{j}_1 = n_1\mathbf{v} + \boldsymbol{\mu} \cdot n_1\mathbf{F}^{\text{ext}} - \mathbf{D} \cdot \nabla n_1, \tag{27}$$

where $\boldsymbol{\mu}$ and $\mathbf{D}$ are the (collective or self) mobility and diffusion matrices that depend nonlinearly on the Peclet number associated with shear.

By symmetry arguments, $\mathbf{X} = \boldsymbol{\mu}, \mathbf{D}$ has the structure:

$$\mathbf{X} = \begin{pmatrix} X_{xx} & X_{xy} & 0 \\ X_{yx} & X_{yy} & 0 \\ 0 & 0 & X_{zz} \end{pmatrix} \tag{28}$$

for the linear flow (9). In general, $\boldsymbol{\mu}$ and $\mathbf{D}$ are nondiagonal in the flow reference frame and asymmetric, unlike the equilibrium case where even in anisotropic media the mobility and diffusion matrices are symmetric due to the Onsager reciprocal relations [61].

Inserting Eq. (27) into (26) shows that evolution of the particle density n_1 depends only on the symmetric part of the tensor $\mathbf{D}$. In principle, however, the antisymmetric part can be determined by a direct measurement of the particle current. It may also influence the behavior of the suspension in the presence of a wall. Similar nondiagonal and asymmetric mobility and diffusion tensors have been encountered in the theory of simple fluids undergoing a strong shear flow (see [62] and references therein).

It is interesting to consider an explicit solution of (26)–(27) for $\mathbf{F}^{\text{ext}} = 0$ and an initial condition in the form of a plane wave, $n_1(\mathbf{r};0) = A_0 \exp(i\mathbf{k}_0 \cdot \mathbf{r})$. For the shear flow (9), the solution has the simple form:

$$n_1(\mathbf{r};0) = A(t)\exp[i\mathbf{k}(t)\cdot\mathbf{r}], \tag{29}$$

with

$$\mathbf{k}(t) = \mathbf{k}_0 - \gamma t(\mathbf{k_0}\cdot\hat{\mathbf{e}}_x)\hat{\mathbf{e}}_y \tag{30}$$

and

$$A(t) = A_0 \exp\left[-\int_0^t dt'\ \mathbf{k}(t')\cdot\mathbf{D}\cdot\mathbf{k}(t')\right]. \tag{31}$$

Several features of this solution are noteworthy. If the initial wave vector $\mathbf{k}_0$ has a nonzero component in the flow direction then, by convection, the component of $\mathbf{k}(t)$ in the gradient direction changes linearly in time. The amplitude $A(t)$ of the density wave decays exponentially. If $\mathbf{k}_0$ is perpendicular to the flow, the exponent of the decay is a linear function of time (as with no flow) and depends on the diffusivity only in the direction of $\mathbf{k}_0$. For $\mathbf{k}_0 \cdot \mathbf{v} \neq 0$, the decay is much faster with the decay-rate being a cubic function of time. Even if $\mathbf{k}_0$ is parallel to the flow direction x, the decay at $t > 0$ depends not only on D_{xx} but also on the diffusion coefficients D_{xy}, D_{yx}, and D_{yy}. The fast decay is associated with Taylor dispersion: as a result of the convection, diffusion across streamlines increases dispersion in the flow direction. (See [57, 63] for the fundamental solution of Eqs. (26)–(28) with $\mathbf{F}^{\text{ext}} = 0$.)

4.3 Observation of Self-Diffusion in a Sheared Suspension

From the solution (29)–(31), it follows that all components of the symmetric part of the self-diffusion tensor can be determined from decay-rate measurements of marked-particle plane density waves. Qiu et al. [60] reported direct experimental observations of the decay of such density patterns in a sheared colloidal suspension using forced Rayleigh scattering technique. Accordingly, colloidal particles were impregnated with a photochromic dye that changes its visible light absorption when illuminated with uv light. Light from a uv laser is split and recombined on the sample producing an interference pattern. Particles photoexcited by a flash of the uv light form a diffraction grating that is observed by probing the sample with a laser beam of visible light.

The observation was performed for 0.073 μm diameter charged polystyrene spheres suspended in deionized water; the volume fraction of solieds was 0.3%. The range of

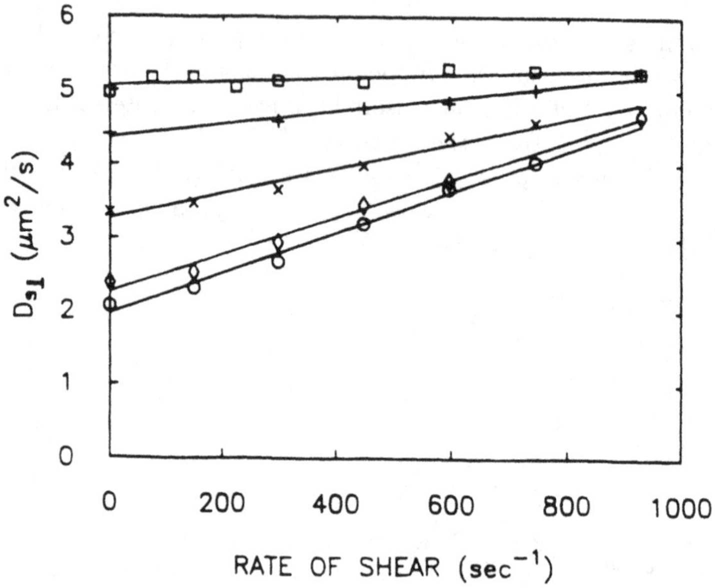

Fig. 11, Transverse self diffusion in charge-stabilized colloidal suspensions in oscillatory shear flow. Different symbols correspond to different strengths of electrostatic repulsion. Systems with stronger interactions have lower values of $D_{s\perp}(\gamma = 0)$ and larger slope. From Qiu et al. [60].

particle electrostatic repulsion was controlled by changing the ionic strength of the solution. At low ionic strengths, the suspension exhibited a behavior characteristic for a concentrated system.

Qiu et al. measured the decay of marked-particle density patterns under oscillatory, rather than stationary shear. The measurements were performed for $Pe = r_{12}^2\gamma/D_s \leq 35$, where D_s is the zero-shear value of the self diffusion coefficient and r_{12} is the average interparticle spacing. The oscillation frequency ω was chosen so that $\gamma/\omega \sim 10$.

The measurements were made for wave-vectors of density patterns in the vorticity and flow directions. Qiu et al. observed the decay of the pattern, averaged over the flow oscillation period. For both directions, they observed an exponential decay $\exp(-D^{\rm eff}k^2t)$, with $D^{\rm eff}$ depending on the shear rate and the direction of $\mathbf{k}$.

Assuming that the evolution of the system is described by a diffusion-convection equation (26)–(28) (with $\mathbf{F}^{\rm ext} = 0$ and oscillatory $\mathbf{v}$), $D^{\rm eff} \equiv D_{s\perp}$ is equal to D_{zz} averaged over the oscillation period for $\mathbf{k}$ in the vorticity (z) direction. Qiu et al. observed an approximately linear dependence of $D^{\rm eff}$ on shear-rate in this case; their results are reproduced in Fig. 11. The dependence of $D_{s\perp} = D_{zz}$ on shear-rate is stronger for more strongly interacting systems.

The longitudinal case, with $\mathbf{k}$ parallel to the flow, is illustrated in Fig. 12. The shear-rate dependence of $D^{\rm eff} \equiv D_{s\|}$ is approximately quadratic since, after averaging

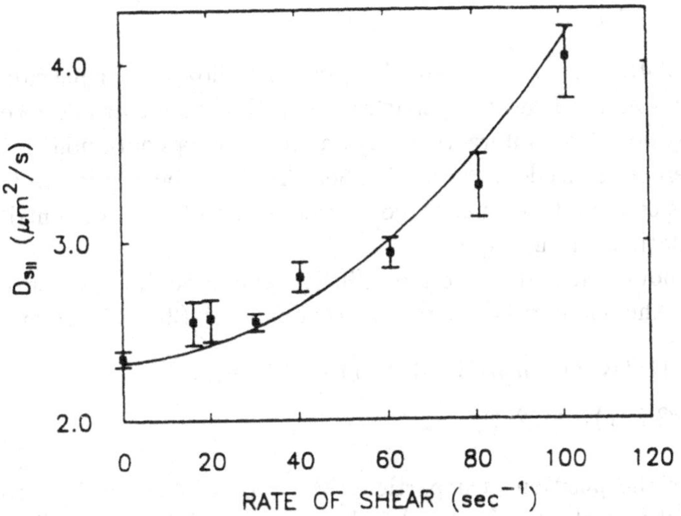

Fig. 12, Longitudinal self diffusion in a charge-stabilized colloidal suspension in oscillatory shear flow. (Dominant quadratic contribution results from Taylor dispersion.) From Qiu et al. [60].

over the oscillation period, the convected wave vector $\mathbf{k}(t)$ produces a $O(\gamma_{av}^2/\omega^2)$ contribution to $D_{s\|}$. Qiu et al. indicated that their results can be approximately described by the expression $D_{s\|} \approx D_{s\perp}(\gamma)(1 + \gamma/2\omega^2)$. For a shear-independent diagonal diffusion matrix, this form implies that D_{xx} and D_{zz} are approximately equal. For the shear dependent diffusion matrix, the results are inconclusive; more detailed measurements are needed to determine D_{xx} and $(D_{xy} + D_{yx})/2$.

Flow-induced particle migration has also been investigated for suspensions of non-Brownian particles. Approximate phenomenological constitutive relations describing shear-induced particle flux have been developed and used to determine particle distributions in suspension flows. For more details regarding this topic see lecture notes by Prof. Schaflinger in this volume.

4.4 Microscopic Analysis of Particle Transport in Shear Flow

Now we present a microscopic analysis of particle migration in shear flow using suspension model similar to that discussed in Sec. 3.4. Evolution equations for the model are introduced in Subsec. 4.4.1. Evolution of small perturbations around the steady state is discussed in Subsec. 4.4.2. The results for the collective mobility and diffusion coefficients are presented in Subsec. 4.4.3 and for the self-mobility and self-diffusion coefficients in Subsec. 4.4.4. The analysis is partly based on Ref. [55] and [56].

4.4.1 Suspension Model

As in Subsec. 3.4.1, we consider a steady plane shear flow of a suspension of identical spherically-symmetric Brownian particles interacting via a pair-additive potential $\Phi(r)$ with no hydrodynamic interactions. An analysis for a continuous interparticle potential is presented; if needed, the hard-sphere limit can be applied to the results. The suspension is semidilute so that three-particle and higher-order contributions in the evolution equations are unimportant.

We consider both self and collective diffusion and mobility problems. On the microscopic level, the one-particle current $\mathbf{j}_1$ [cf. Eqs. (26)–(27)] can be expressed as:

$$\mathbf{j}_1(1;t) = n_1(1;t)\mathbf{v}(1) + \mu_0 n_1(1;t)\mathbf{F}^{\text{ext}}(1) - D_0\nabla_1 n_1(1;t) + \mu_0 \int d2\; \mathbf{F}(\mathbf{r})\, n_2(1,2;t), \tag{32}$$

where $(i) = \mathbf{r}_i$ is the position of a particle, $\nabla_1 = \partial/\partial\mathbf{r}_1$, D_0 and μ_0 are the diffusion and mobility constants of an isolated particle, and $\mathbf{F}(\mathbf{r}) = -\nabla_1\Phi(r)$ is the interparticle force.

Symbols $\mathbf{j}_1$, n_1, and $\mathbf{F}^{\text{ext}}$ for the self and collective problems are defined at the beginning of Sec. 4.2. For the self problem, the variable (1) refers to the tagged particles and the variable (2) to the untagged ones. For example, $n_1(1;t) \equiv n_s(1;t)$ denotes the density of tagged particles, while $n_1(2;t) \equiv n$ denotes the uniform density of untagged particles. For the collective problem, all particles are equivalent as are variables (1) and (2).

For a semidilute suspension, the evolution equation for the two-particle density n_2 is closed:

$$\left[\frac{\partial}{\partial t} - \hat{\Omega}_c(1,2) - \hat{E}(1,2)\right] n_2(1,2;t) = 0, \tag{33}$$

where $\hat{\Omega}_c(1,2)$ denotes the diffusion-convection operator for two interacting particles,

$$\hat{\Omega}_c(1,2) = \hat{\Omega}(1,2) + \gamma\hat{C}(1,2), \tag{34}$$

with $\hat{\Omega}$ denoting the two-particle Smoluchowski operator,

$$\hat{\Omega} = -\sum_{i=1}^{2} \nabla_i \cdot [-D_0\nabla_i - \mu_0\nabla_i\Phi(r)], \tag{35}$$

and $\hat{C}$ describing the contribution due to the convective flow,

$$\gamma\hat{C} = -\sum_{i=1}^{2} \nabla_i \cdot \mathbf{v}(\mathbf{r}_i). \tag{36}$$

The operator $\hat{E}$ represents the influence of the external force. For the collective problem,

$$\hat{E}(1,2) = -\mu_0 \sum_{i=1}^{2} \nabla_i \cdot \mathbf{F}^{\text{ext}}(i), \tag{37}$$

and for the self problem,

$$\hat{E}(1,2) = -\mu_0 \nabla_1 \cdot \mathbf{F}^{\text{ext}}(1). \tag{38}$$

4.4.2 Decomposition of Pair Density

Our purpose is to derive from Eqs. (32)–(33) a closed linear constitutive equation of the form (27). Thus, we need to evaluate n_2 from the two-particle equation (33). It is convenient to divide n_2 into two parts,

$$n_2(1,2;t) = n_1(1;t)n_1(2;t)g^{\text{st}}(\mathbf{r}) + \delta n_2(1,2;t), \tag{39}$$

where g^{st} is the low-density steady state correlation in a uniform system with no external force (cf. Sec. 3.4). The first term describes the steady-state-like portion of the correlations. At $t = 0$, we assume that $\delta n_2 = 0$.

The evolution equation for δn_2 is obtained by inserting (39) into the two-particle equation (33) and using the steady-state equation

$$\hat{\Omega}_c g^{\text{st}} = 0. \tag{40}$$

After some transformations, we find:

$$\left[\frac{\partial}{\partial t} - \hat{\Omega}_c(1,2)\right] \delta n_2(1,2;t) = -\mu_0[1 + P(12)]n_1(1;t)n_1(2;t)\mathbf{s}_1(\mathbf{r})\mathbf{F}^{\text{ext}}(1)$$
$$-\mu_0[1 + P(12)]n_1(2;t)\mathbf{s}_2(\mathbf{r})[-k_B T \nabla_1 n_1(1;t)], \tag{41}$$

where

$$\mathbf{s}_1(\mathbf{r}) = \nabla_1 g^{\text{st}}(\mathbf{r}), \tag{42}$$

and

$$\mathbf{s}_2(\mathbf{r}) = 2\nabla_1 g^{\text{st}}(\mathbf{r}) - \beta \mathbf{F}(\mathbf{r}) g^{\text{st}}(\mathbf{r}). \tag{43}$$

For the collective problem $P(12)$ is the particle permutation operator:

$$P(12)f(1,2) = f(2,1), \tag{44}$$

and for the self problem:

$$P(12) = 0. \tag{45}$$

In Eq. (41), only lowest-order terms in density have been retained; and the external force term of the evolution operator has been neglected because only the linear transport coefficients are sought.

The macroscopic constitutive equation (27) is obtained by inserting the solution of Eq. (41) into (32). On the macroscopic level, time scales are much longer than the Brownian relaxation time τ_B, and the fields $\mathbf{F}^{\text{ext}}$ and n_1 vary on a much larger length-scale l than the interaction range b. Therefore, only the long-time, long-wavelength limit of Eq. (41) is required.

The appropriate analysis is presented in Subsec. 4.4.3 for the collective problem and in Subsec. 4.4.4 for the self problem.

4.4.3 Collective Mobility and Diffusion

Let us consider the long-wavelength limit of Eq. (41) in the collective problem. For a constant force $\mathbf{F}^{\text{ext}}$ and density gradient n_1, the right side vanishes by antisymmetry of the source terms $\mathbf{s}_1$ and $\mathbf{s}_2$. Thus,

$$\delta n_2(1,2;t) \equiv 0, \tag{46}$$

so that correlations are unperturbed. This simplification follows because a constant force on all particles does not change their relative positions.[1]

Taking into account Eq. (46), the constitutive relation (27) is obtained by inserting (39) into (32) and expanding $n_1(2;t)$ around $\mathbf{r}_1$. By retaining only the leading terms, (27) is obtained with the collective mobility and diffusion tensors $\boldsymbol{\mu}$ and $\mathbf{D}$ given by:

$$\boldsymbol{\mu} = \mu_0 \mathbf{I} \tag{47}$$

and

$$\mathbf{D} = \mu_0 \left[k_B T \mathbf{I} - n_1 \int d^3r \; \mathbf{r} \mathbf{F}(\mathbf{r}) g^{\text{st}}(\mathbf{r}) \right] \tag{48}$$

The bracketed expression in Eq. (48) can be expressed in terms of the particle contribution $\boldsymbol{\sigma}$ to the stress tensor at the steady state. Using the standard expression for the stress tensor in a colloidal suspension without hydrodynamic interactions [4], the following relation between the tensors $\boldsymbol{\mu}$ and $\mathbf{D}$ is obtained:

$$\mathbf{D} = -\boldsymbol{\mu} \cdot \frac{\partial \boldsymbol{\sigma}}{\partial n}, \tag{49}$$

where the partial derivative is evaluated at constant shear rate and temperature.

Equation (49) is a natural generalization of the Einstein relation for equilibrium systems [4] with $-\boldsymbol{\sigma}$ replacing the osmotic pressure. However, this result was obtained for a specific system; it may not be generally valid.

[1] Indeed, for monodisperse suspensions with no hydrodynamic interactions, a similar property holds at arbitrary densities. For a similar reason, this property also holds for suspensions with hydrodynamic interactions but only at the two-particle level.

Using Eq. (49) and the results of Subsec. 3.4.4 (supplemented by the corresponding result for the isotropic part of the stress tensor not shown), one can explicitly evaluate the collective diffusion tensor $\mathbf{D}$ for a hard-sphere potential. For a semidilute suspension of non-Brownian particles, the results presented in Subsec. 3.4.5 apply.

4.4.4 Self-Mobility and Self-Diffusion

The self-mobility and self-diffusion problems are more complex because the source term at the right hand side of Eq. (41) is nonzero, even in the long-wave limit. Since we are considering the linear transport processes, solutions of Eq. (41) for the self-mobility and self-diffusion are discussed separately.

For self-mobility the system is statistically uniform, and a constant external force $\mathbf{F}^{\text{ext}}$ acts on tagged particles. In order to obtain the constitutive relation (27), equation (41) is formally solved and the limiting long-time solution is inserted into (32). The resulting expression for the self-mobility tensor $\boldsymbol{\mu}_s \equiv \boldsymbol{\mu}$ can be written as [55, 56]:

$$\boldsymbol{\mu}_s = \mu_0 \left[\mathbf{I} + 2\phi\mathbf{M}\right], \tag{50}$$

where

$$\mathbf{M} = \frac{3}{8\pi b^3} \int d^3r \; \mathbf{F}(\mathbf{r}) \hat{\Omega}_c^{-1} \mathbf{s}_1(\mathbf{r}), \tag{51}$$

and $\phi = \frac{4}{3}\pi b^3$.

Before presenting explicit results for the self-mobility, we derive the formal expression for the self-diffusion tensor. Accordingly, we set $\mathbf{F}^{\text{ext}} = 0$ and consider a tagged-particle density wave. The wave-vector of the density wave evolves due to convection; thus, we seek a solution to the microscopic equations (26), (32), (39), and (41) with $n_1(1;t) \equiv n_s(1;t)$ in the form of (29)–(30). One can check *a posteriori* that a solution of this form exists and, in the long-wave limit, $A(t)$ is a slowly varying function of time.

Next, we factor δn_2:

$$\delta n_2(1, 2, ; t) = H(\mathbf{r}; t) \exp[i\mathbf{k}(t) \cdot \mathbf{R}], \tag{52}$$

where $\mathbf{r}$ is the relative particle position and $\mathbf{R} = \frac{1}{2}(\mathbf{r}_1 + \mathbf{r}_2)$ is the center of mass. After inserting (52) into (41) and retaining only the leading term in $\mathbf{k}(t)$, we obtain:

$$\left[\frac{\partial}{\partial t} - \hat{\Omega}_c\right] H(\mathbf{r}; t) = -iD_0\mathbf{k}(t)A(t)n\mathbf{s}_2(\mathbf{r}). \tag{53}$$

The equation (53) is formally solved with a constant A (for small $\mathbf{k}$ slowly varying on the time scale τ_B), the result is inserted into (32), and the long-time, long-wavelength limit is taken. To obtain the limiting form, it is important to recall that $\mathbf{k}(t)$ grows linearly with time for $\mathbf{k}_0 \cdot \mathbf{v} \neq 0$ [cf. Eq. (30)]. Therefore, we must first take the limit $\mathbf{k}_0 \to 0$ and then $t \to \infty$. Physically, it means that even for small k_0, the range of validity of the constitutive equation (27) is limited to a finite period $\tau_b \ll t \ll (\gamma k_{0x} b)^{-1}$. The resulting self-diffusion tensor $\mathbf{D}_s = \mathbf{D}$ is:

$$\mathbf{D}_s = D_0 \left[\mathbf{I} + 2\phi(\mathbf{M}_2 + \mathbf{M}_1)\right] \tag{54}$$

with

$$\mathbf{M}_1 = \frac{3}{8\pi b^3} \int d^3r \; \mathbf{F}(\mathbf{r}) \hat{\Omega}_c^{-1} \mathbf{s}_2(\mathbf{r}) \tag{55}$$

and

$$\mathbf{M}_2 = \frac{3\gamma}{8\pi b^3} \int d^3r \; \mathbf{F}(\mathbf{r}) \, \hat{\mathbf{e}}_x \int_0^\infty d\tau \; \tau \; \exp(\hat{\Omega}_c \tau) s_{2y}(\mathbf{r}), \tag{56}$$

where s_{2y} is the y component of $\mathbf{s}_2$ and $\mathbf{M}_2$ results from the time-dependence of $\mathbf{k}$.

The foregoing derivation of the self-diffusion tensor $\mathbf{D}_s$ is based on ideas in Ref. [55]; there, however, $\mathbf{M}_2$ was mistakenly omitted.

Let us compare the expressions for the self-mobility and self-diffusion matrices. In the absence of shear, $\mathbf{M}_1 = \mathbf{M}$ and $\mathbf{M}_2 = 0$. [The former relation is obtained by noting that $g^{st} = \exp(-\beta\Phi)$ and thus, $\mathbf{s}_1 = \mathbf{s}_2$ as follows from (42)–(43); the latter follows from (56).] Hence, our results are consistent with the Einstein relation $D_s = k_B T \mu_s$ between the equilibrium self-mobility and self diffusion constants. Out of equilibrium, however, $\mathbf{s}_1 \neq \mathbf{s}_2$ and $\mathbf{M}_2 \neq 0$. Therefore, in nonequilibrium steady states we usually have $\mathbf{D}_s \neq k_B T \boldsymbol{\mu}_s$, so the Einstein relation between the self-mobility matrix $\boldsymbol{\mu}$ and the self-diffusion matrix $\mathbf{D}$ is not valid.

We now present several explicit results. The $O(Pe)$ behavior of the self-mobility and self-diffusion tensors can be obtained using the regular perturbation in Pe [55]. Beyond the linear term, the regular expansion in powers of Pe does not exist but an expansion in powers of $Pe^{1/2}$ can be obtained [56]. For hard spheres, the leading-order terms were calculated analytically. For $\mathbf{M}$ and $\mathbf{M}_1$ one gets [55, 56]:

$$\overleftrightarrow{M}(Pe) = \begin{pmatrix} -1 & -\frac{4}{15}Pe & 0 \\ \frac{11}{15}Pe & -1 & 0 \\ 0 & 0 & -1 \end{pmatrix} + O(Pe^{3/2}). \tag{57}$$

$$\mathbf{M}_1 = \mathbf{M} + o(Pe). \tag{58}$$

And for $\mathbf{M}_2$ the result is:

$$\mathbf{M}_2 = Pe \, \hat{\mathbf{e}}_y \hat{\mathbf{e}}_x + o(Pe). \tag{59}$$

The leading corrections $\mathbf{X} = X^{(0)}\mathbf{I} + Pe\,\mathbf{X}^{(1)} + o(Pe)$ to $\mathbf{X} = \mathbf{M}, \mathbf{M}_1$ and $\mathbf{M}_2$ were calculated numerically for the Yukawa potential (6). The results are plotted as functions of the hardness parameter α in Fig. 13 for $\mathbf{M}$ and $\mathbf{M}_1$ and Fig. 14 for $\mathbf{M}_2$. Note that the relation(58) is not valid for the Yukawa potential.

For hard spheres, the matrix $\mathbf{M}$ was evaluated for small and intermediate Peclet numbers by a $Pe^{1/2}$-expansion method [56]. The Peclet-number dependence of various

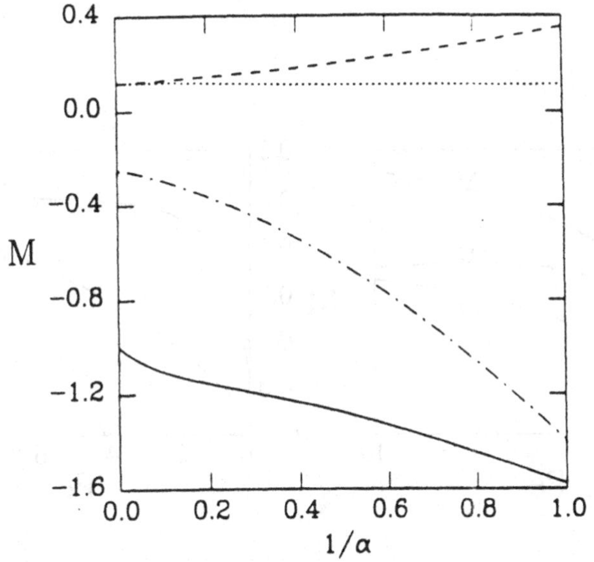

Fig. 13, Self-mobility (50) and self-diffusion (54) tensors for sheared semidilute suspension of charged particles interacting via screened Coulombic potential (6) as functions of the inverse hardness parameter $1/\alpha$. The $O(1)$ contributions: $M^{(0)} \equiv M_1^{(0)}$ (solid line). The $O(Pe)$ contributions: symmetric part of $\mathbf{M}^{(1)}/|M^{(0)}|$ (dashed line), symmetric part of $\mathbf{M}_1^{(1)}/|M^{(0)}|$ (dotted line), antisymmetric part of $\mathbf{M}^{(1)}/|M^{(0)}| = \mathbf{M}_1^{(1)}/|M^{(0)}|$ (dash-dotted line). From Szamel et al. [55].

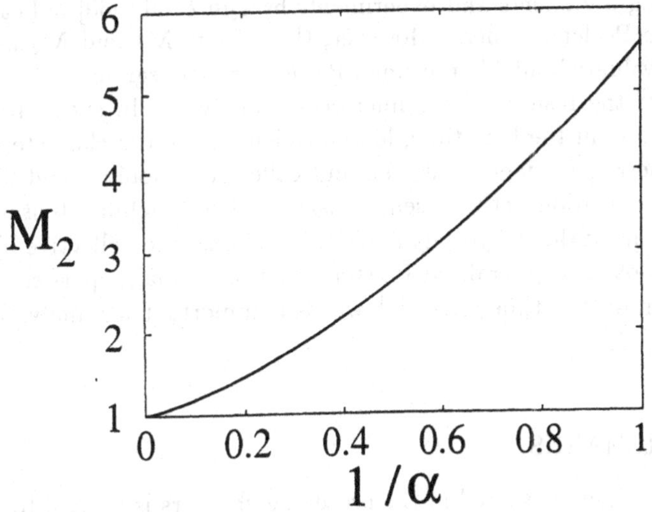

Fig. 14, Nonzero component of $O(Pe)$ contribution $\mathbf{M}_2^{(1)}/|M^{(0)}|$ to self-diffusion tensor (54) for sheared semidilute suspension of charged particles interacting via screened Coulombic potential (6) as functions of the inverse hardness parameter $1/\alpha$.

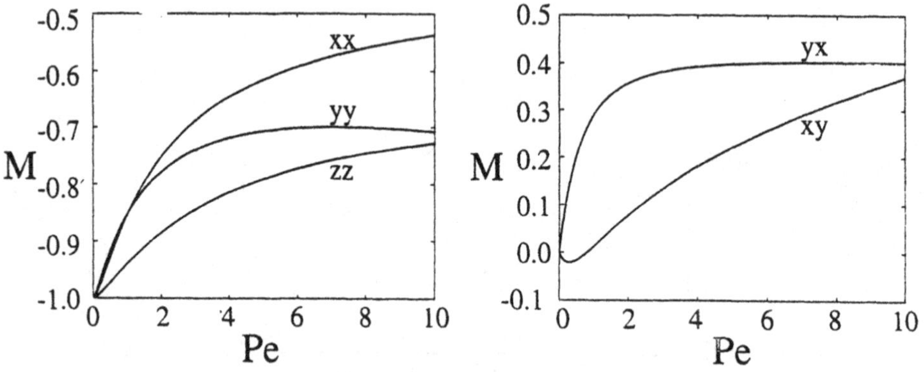

Fig. 15, Diagonal (a) and off-diagonal (b) components of **M**, the two-particle contribution to self-mobility tensor, as functions of *Pe*. Corresponding matrix components are indicated.

components of the mobility tensor **M** is presented in Fig. 15. Note that the linear approximation (57) is valid only for $Pe < 0.02$ (a range hardly visible on the scale of the figure). Scaling arguments and experiments by Qiu et al. [60] indicate that $\mathbf{D}/D_0 \sim Pe$ at large Peclet numbers. However, the tensors $\mathbf{M}_1$ and $\mathbf{M}_2$ have not been yet explicitly evaluated outside the small Peclet-number regime.

Let us summarize the results of our microscopic analysis. In sheared suspensions, particle migration in the long-time, long-wavelength limit is characterized by nondiagonal, asymmetric, *Pe*-dependent (self and collective) mobility and diffusion tensors. The Einstein relation between self-diffusion and self-mobility tensors is violated. For suspensions without hydrodynamic interactions, the collective diffusion and mobility tensors obey a generalized Einstein relation. Finally, particle density evolves by convection on the time scale γ^{-1} which is important for analysis of the diffusion process.

5. CONCLUDING REMARKS

The theoretical analysis presented in the preceding chapters is limited to semidilute suspensions. The experiments, however, are performed at moderate and high effective volume fractions (based on the interaction range) because structure factor and the effect of the flow on transport properties are difficult to measure at low concentrations. As discussed in Sec. 3.4, the mode-mode coupling approximation, used to describe dense suspensions, does not yield quantitative results.

For dense hard-sphere suspensions close to equilibrium, good approximations have recently been proposed by Cichocki et al. [64, 65] and by Leegwater and Szamel [66, 67]. The approximations are derived from the Enskog concept (in the theory of simple fluids) that evolution of a dense system can be described in terms of uncorrelated binary collisions whose frequency is renormalized by the equilibrium correlations. Cichocki proposed [68] an approximation scheme where the diffusion of two interacting Brownian spheres corresponds to a binary collision in the Enskog theory. The contact Enskog approximation [65, 66], using the contact value of the pair correlation function to renormalize the frequency of two-particle encounters, is the simplest accurate application of the Enskog concept to colloidal suspensions. It produces good results up to a volume fraction $\phi = 0.4$. Mode-mode coupling contributions, important for higher volume fractions, were incorporated in the Enskog-like theories [67, 64]. We expect that application of these ideas in sheared suspension theories will also yield good results.

The theoretical analysis presented in these lectures neglects hydrodynamic interactions. While this is a reasonable approximation for strongly charged colloidal particles suspended in weakly ionic solvents, it is insufficient for other systems. A full analysis of the properties of sheared colloidal suspensions with hydrodynamic interactions is yet to be accomplished; however, there has been a recent progress in this direction.

Brady and his collaborators [53, 59] analyzed the two particle Smoluchowski equation for a hard-sphere suspension with hydrodynamic interactions at moderate Peclet numbers. They have evaluated the structure factor and the stress tensor up to the second order in Pe.

Zinchenko recently developed an efficient algorithm capable of solving the two particle Smoluchowski equation with hydrodynamic interactions for arbitrary Peclet numbers (Zinchenko and Davis [69]). Using this method, Zinchenko [54] has calculated the stress tensor for a sheared semidilute hard sphere suspension with hydrodynamic interactions. He observed that with increasing Peclet number, the viscosity first decreases then slightly increases. Shear thinning followed by slight shear thickening has been experimentally observed [42].

Finally, there are similarities between colloidal suspensions and simple fluids. Shear-rate dependent nonequilibrium steady states in fluids have been studied extensively [62], [70]–[72]. In simple fluids, the structural relaxation time is 10^{-11}s. Shear rates corresponding to $Pe \gtrsim 1$ are very difficult to obtain in experiments and therefore, shear induced nonequilibrium steady states have been investigated mostly theoretically and by computer simulations. However, nonlinear distortion of the structure factor has been observed in nonequilibrium steady states generated by a large temperature gradient [73].

ACKNOWLEDGEMENTS

I thank Prof. M. Loewenberg for valuable discussions and his careful reading of the manuscript, and acknowledge Department of Chemical Engineering at Yale University for its hospitality.

REFERENCES

1. Beysens, D., M. Gbadamassi, and L. Boyer: Light-scattering study of a critical mixture with shear flow, Phys. Rev. Lett., 43 (1979), 1253–1255.
2. Klein, R.: The structure and dynamics of strongly interacting charged colloidal liquids., in: Structure and Dynamics of Strongly Interacting Colloids and Supramolecular Aggregates in Solution (Ed. S.-H. Chen, J. S. Huang and P. Tartaglia), Kluwer Academic Publishers, Dordrecht 1992.
3. Russel, W. B.: The dynamics of colloidal systems, University of Wisconsin Press, London 1987.
4. Russel, W. B., D. A. S. Saville, and W. R. Schowalter: Colloidal dispersions, Cambridge University Press, Cambridge 1989.
5. Pusey, P. N.: Colloidal suspensions, in: Liquids, Freezing and the Glass Transition (Ed. D. Levesque, J.-P. Hansen, and J. Zinn-Justin), Elsevier, Amsterdam 1990.
6. Kim, S. and S. J. Karrila: Microhydrodynamics: Principles and selected applications, Butterworth-Heinemann, London 1991.
7. Durlofsky, L., J. F. Brady, and G. Bossis: Dynamic simulations of hydrodynamically interacting particles, J. Fluid. Mech., 180 (1987), 21–49.
8. Brady, J. F. and G. Bossis: Stokesian dynamics, Ann. Rev. Fluid Mech., 20 (1988), 111–157.
9. Ladd, A. J. C.: Hydrodynamic interactions in a suspension of spherical particles, J. Chem. Phys., 88 (1988), 5051–5063.
10. Ladd, A. J. C.: Hydrodynamic interactions and the viscosity of suspensions of freely moving spheres, J. Chem. Phys., 90 (1989), 1149–1157.
11. Ladd, A. J. C.: Hydrodynamic transport coefficients of random dispersion of hard spheres, J. Chem. Phys., 93 (1990), 3484–3494.
12. Cichocki, B., B. U. Felderhof, K. Hinsen, E. Wajnryb, and J. Bławzdziewicz: Friction and mobility of many spheres in Stokes flow, J. Chem. Phys., 100 (1994), 3780–3790.
13. Cichocki, B. and K. Hinsen: Stokes drag on conglomerates of spheres, Phys. Fluids, 7 (1995), 285–291.
14. Ladd, A. J. C.: Numerical simulations of particulate suspensions via a discretized Boltzmann equation. Part 1. Theoretical foundation, J. Fluid Mech., 271 (1994), 285–309.

15. Ladd, A. J. C.: Numerical simulations of particulate suspensions via a discretized Boltzmann equation. Part 2. Numerical results, J. Fluid Mech., 271 (1994), 311–339.
16. Felderhof, B. U.: Sedimentation and convective flow in suspensions of spherical particles, Physica A, 153 (1988), 217–233.
17. Felderhof, B. U.: Dynamics of hard sphere suspensions, Physica A, 169 (1990), 1–16.
18. Hansen, J. P. and I. R. McDonald: Theory of simple liquids, Academic Press, London 1976.
19. Nägele, G., M. M. Medina-Noyola, R. Klein, and J. L. Arauz-Lara: Time-dependent self-diffusion in model suspension of highly charged Brownian particles, Physica A, 149 (1988), 123–163.
20. Hoffman, R. L.: Discontinuous and dilatant viscosity behavior in concentrated suspensions. I. Observations of a flow instability, Trans. Soc. Rheol., 16 (1972), 155–173.
21. Clark, N. A. and B. J. Ackerson: Observations of the coupling of concentration fluctuations to steady-state shear flow, Phys. Rev. Lett., 44 (1980), 1005–1008.
22. Ackerson, B. J.: Shear induced order and shear processing of model hard sphere suspension, J. Rheol., 34 (1990), 553–590.
23. Ackerson, B. J.: Shear induced order in equilibrium colloidal liquids, Physica A, 174 (1991), 15–30.
24. Chen, L. B., C. F. Zukoski, B. J. Ackerson, H. J. M. Hanley, G. C. Straty, J. Barker, and C. J. Glinka: Structural changes and orientational order in a sheared colloidal suspension, Phys. Rev. Lett., 69 (1992), 688–691.
25. Ackerson, B. J., J. van der Werff, and C. G. de Kruif: Hard-sphere dispersions: Small-wave-vector structure-factor measurements in a linear shear flow, Phys. Rev. A, 37 (1988), 4819–4827.
26. Johanson, S. J., C. G. de Kruif, and R. P. May: Structure factor distortion for hard-sphere dispersions subjected to weak shear flow: Small-angle neutron scattering in the flow-vorticity plane, J. Chem. Phys., 89 (1988), 5909–5921.
27. van der Werff, J. C., B. J. Ackerson, R. P. May, and C. G. de Kruif: Neutron scattering from dense colloidal dispersions at high shear rates: The deformation of the structure factor in the shear plane, Physica A, 195 (1990), 375–398.
28. Wagner, N. J. and W. B. Russel: Light scattering measurements of a hard-sphere suspension under shear, Phys. Fluids A, 2 (1990), 491–502.
29. Woutersen, A. T. J. M, R. P. May, and C. G. de Kruif: The shear-distorted microstructure of adhesive hard sphere dispersions: A small-angle neutron scattering study, J. Rheol., 37 (1993), 71–88.
30. Yan, Y. D. and J. K. Dhont: Shear-induced structure distortion in nonaqueous dispersions of charged colloidal spheres via light scattering, Physica A, 198 (1993), 78–107.

31. Chow, M. K. and C. F. Zukoski: Nonequilibrium behavior of dense suspensions of uniform particles: Volume fraction and size dependence of rheology and microstructure, J. Rheology, 39 (1995), 33–58.
32. Ackerson, B. J., J. B. Hayter, N. A. Clark, and L. Cotter: Neutron scattering from charge stabilized suspensions undergoing shear, J. Chem. Phys., 84 (1986), 2344–2349.
33. Ashdown, S., I. Marković, R. H. Ottewill, P. Lindner, R. C. Oberthür, and A. R. Rennie: Small-angle neutron scattering studies on ordered polymer colloid dispersion, Langmuir, 6 (1990), 303–307.
34. Evans, D. J., S. T. Cui, H. J. M. Hanley, and G. C. Straty: Conditions for the existence of a reentrant solid phase in a sheared atomic fluid, Phys. Rev. A, 46 (1992), 6731–6734.
35. Ackerson, B. J. and P. N. Pusey: Shear induced order in suspensions of hard spheres, Phys. Rev. Lett., 61 (1988), 1033–1036.
36. Yan, Y. D., J. K. G. Dhont, C. Smits, and H. N. W. Lekkerkerker: Oscillatory-shear-induced order in nonaqueous dispersions of charged colloidal spheres, Physica A, 202 (1994), 68–80.
37. Bossis, G., Y. Grasseli, E. Lemaire, A. Meunier, J. F. Brady, and T. Phung: Rheology and microstructure in colloidal suspensions, Phys. Scripta, T49 (1993), 89–93.
38. Heyes, D. M.: Shear thinning of dense suspensions modelled by Brownian dynamics, Phys. Lett. A, 132 (1988), 399–402.
39. Xue, W. and G. S. Grest: Shear-induced alignment of colloidal particles in the presence of a shear flow, Phys. Rev. Lett., 64 (1990), 419–422.
40. Krieger, I. M.: Rheology of monodisperse lattices, Adv. Colloid Interface Sci., 3 (1972), 111–136.
41. Russel, W. B.: Effects of interactions between particles on the rheology of dispersions, in: Theory of Dispersed Multiphase Flow (Ed. R. E. Mayer), Academic Press, New York 1983.
42. Choi, G. N. and I. M. Krieger: Rheological studies on sterically stabilized model dispersions of uniform colloidal spheres. II. Steady-shear viscosity, J. Colloid Interface Sci., 113 (1986), 101–113.
43. van der Werff, J. C. and C. G. de Kruif: Hard-sphere colloidal dispersions: The scaling of rheological properties with particle size, volume fraction, and shear rate, J. Rheol., 33 (1989), 421–454.
44. van der Werff, J. C., C. G. de Kruif, and J. K. G. Dhont: The shear-thinning behaviour of colloidal dispersions. II. Experiments, Physica A, 160 (1989), 205–212.
45. Chow, M. K. and C. F. Zukoski: Gap size and shear history dependencies in shear thickening of a suspension ordered at rest, J. Rheology, 39 (1995), 15–32.

46. Al-Hadithi, T. S. R., H. A. Barnes, and K. Walters: The relationship between the linear (oscillatory) and nonlinear (steady-state) flow properties of a series of polymer and colloidal systems, Colloid Polym. Sci., 270 (1992), 40–46.
47. Johma, A. I. and P. A. Reynolds: An experimental study of the first normal stress difference—shear stress relationship in simple shear flow for concentrated shear thickening suspensions, Rheol. Acta, 32 (1993), 457–464.
48. Laun, J. M.: Normal stresses in extremely shear thickening polymer dispersions, J. Non-Newt. Fluid Mech., 54 (1994), 87–108.
49. Ronis, D.: Theory of fluctuations in colloidal suspensions undergoing steady shear flow, Phys. Rev. A, 29 (1984), 1453–1460.
50. Schwarzl, J. F. and S. Hess: Shear-flow-induced distortion of the structure of a fluid: Application of a simple kinetic equation, Phys. Rev. A, 33 (1986), 4277–4283.
51. Dhont, J. K. G.: On the distortion of the static structure factor of colloidal fluids in shear flow, J. Fluid Mech., 204 (1989), 421–431.
52. Bławzdziewicz, J. and G. Szamel: Structure and rheology of semidilute suspension under shear, Phys. Rev. E, 48 (1993), 4632–4636.
53. Brady, J. F. and M. Vicic: Normal stresses in colloidal dispersions, J. Rheology, 39 (1995), 545–566.
54. Zinchenko, A. Z.: Private communication (1995).
55. Szamel, G., J. Bławzdziewicz, and J. A. L. Leegwater: Self-diffusion in sheared suspensions: Violation of Einstein relation, Phys. Rev. A, 45 (1992), R2173–R2176.
56. Bławzdziewicz, J. and M. Ekiel-Jezewska: How shear flow of a semidilute suspension modifies its self-mobility, Phys. Rev. E, 51 (1995), 4704–4708.
57. Erlick, D. E.: Source functions for diffusion in uniform shear flow, Aust. J. Phys., 15 (1962), 283–288.
58. Hess, S.: Similarities and differences in the nonlinear flow behavior of simple and molecular fluids, Physica A, 118 (1983), 79–104.
59. Brady, J. F. and J. F. Morris: Microstructure in a strongly-sheared suspension and its impact on rheology and diffusivity, in: IUTAM Symposium on Hydrodynamic Diffusion of Suspended Particles, Estes Park, Colorado, 22-25 July 1995.
60. Qiu, X., H. D. Ou-Yang, D. J. Pine, and Chaikin P. M.: Self-diffusion of interacting colloids far from equilibrium, Phys. Rev. Lett., 61 (1988), 2554–2557.
61. de Groot, S. R. and P. Mazur: Nonequilibrium thermodynamics, Dover, New York 1984.
62. Sarman, S., D. J. Evans, and A. Baranyai: Mutual and self-diffusion in fluids undergoing strong shear, Phys. Review A, 46 (1992), 893–902.
63. Sarman, S., D. J. Evans, and P. T. Cummings: Comment on: Nonequilibrium molecular dynamics calculation of self-diffusion in a non-Newtonian fluid subject to a Couette strain field, J. Chem. Phys., 95 (1991), 8675–8676.

64. Cichocki, B. and K. Hinsen: Self and collective diffusion coefficients of hard sphere suspensions, Ber. Bunsenges. Phys. Chem., 94 (1990), 243–246.
65. Cichocki, B. and B. U. Felderhof: Dynamic scattering function of dense suspension of hard spheres, Physica A, 204 (1994), 152–168.
66. Leegwater, J. A. L. and G. Szamel: Dynamical properties of hard-sphere suspensions, Phys. Rev. A, 46 (1992), 4999–5011.
67. Szamel, G. and J. A. L. Leegwater: Long-time self-diffusion coefficients of suspensions, Phys. Rev. A, 46 (1992), 5012–5018.
68. Cichocki, B.: Linear kinetic theory of a suspension of interacting Brownian particles. II. Density-density autocorrelation function, Physica A, 148 (1988), 191–207.
69. Zinchenko, A. Z. and R. H. Davis: Gravity induced coalescence of drops at arbitrary Péclet numbers, J. Fluid Mech., 280 (1994), 119–148.
70. Daivis, P. J. and D. J. Evans: Thermal conductivity of shearing fluid, Phys. Rev. A, 48 (1993), 1058–1065.
71. Schmitz, R.: Fluctuations in nonequilibrium fluids, Phys. Rep., 171 (1988), 1–58.
72. Kirkpatrick, T. R. and J. C. Nieuwoudt: Stability of dense simple fluids subjected to large shear: Shear induced ordering, Phys. Rev. A, 40 (1989), 5238–5248.
73. Segre, P. N., R. W. Gammon, and J. V. Sengers: Light-scattering measurements of nonequilibrium fluctuations in liquid mixture, Phys. Rev. A, 47 (1993), 1026–1034.

THE MACROSCOPIC MODELLING OF MULTI-PHASE MIXTURES

D. Lhuillier
Pierre et Marie Curie University, Paris, France

Abstract

The usual classification of materials into solids, liquids and gases is not entirely satisfying. Those who already used a tooth-paste, made a mayonnaise or wiped muddy shoes have experienced materials with strange properties, neither completely solid nor completely liquid. There also exist liquid-like materials (paint, egg white etc...) which do not behave as ordinary liquids. A careful investigation of those soft solids or odd liquids reveals they are all made of several chemical species and that the basic blocks of material are of a supramolecular size. In fact, these complex fluids are multi-phase mixtures, the simplest example of which is a suspension of particles in a fluid .The present lectures will concern the macroscopic modelling of multi-phase mixtures. Let us insist on the two underlined terms. Macroscopic: we will not focuss on the way to get exact results concerning the flow around particles but we will select among these results, the ones which prove important when considering the mixture as a continuous medium. Modelling: the description to be given will be a simplified one, obtained after a selection of the relevant physical phenomena, and a choice of the best variables to represent them. In the first chapter we will present fluid-mechanical results, while in chapter two we will consider the possible implications of the second law of thermodynamics. In these two chapters, a certain number of (more or less) intuitive statements will be made concerning the averaging procedure. These statements will be justified in chapter three while the last chapter will insist on phenomena linked with the fluctuational kinetic energy.

NOTATION

e_n°, e_n, e	: total energy per unit mass (3.5)
$\mathbf{g}_n^{\circ}$, $\mathbf{g}_n$, $\mathbf{g}$	: external force per unit mass (3.3)
$\mathbf{h}_n^{\circ}$, $\mathbf{h}_n$, $\mathbf{h}$	: entropy flux (3.6)
p_n°, p_n, p	: pressure (3.6)
$\mathbf{q}_n^{\circ}$, $\mathbf{q}_n$, $\mathbf{q}$	: heat flux (3.5)
s_n°, s_n, s	: entropy per unit mass (3.6)
T_n°, T_n, T	: temperature (3.6)
$\mathbf{v}_n^{\circ}$, $\mathbf{v}_n$, $\mathbf{v}$	: mass-weighted velocity (3.2)
Δ_n°, Δ_n, Δ	: entropy production rate per unit volume (3.6)
ε_n°, ε_n, ε	: internal energy per unit mass (3.5)
ρ_n°, ρ_n, ρ	: mass per unit volume (3.2)
σ_n°, σ_n, σ	: stress (3.3)
τ_n°, τ_n, τ	: deviatoric stress (3.6)

For all the above quantities, X_n° is a local-instant value, X_n the average for phase n and X the average for the whole mixture. The index n has two values for a two-phase mixture, either p (particles) and f (fluid), or 1 (fluid) and 2 (particles) .

a	: radius of a spherical particle
c_n	: mass fraction of phase n (c stands for c_p) (2.2)
D	: coefficient of concentration-diffusion (2.2)
E_n	: exchange of total energy (3.5)
E_n^*	: exchange of internal energy (3.5)
G_n , G^*	: power developped by fluctuations of the external forces (3.5)
g, Δg_n, Δg, Δg^*	: free-enthalpies (2.2) and (2.3)
$h(\phi)$	: hindrance function (1.3)
$\mathbf{H}$	: total entropy flux (3.6)
$\mathbf{J}_n$	: relative mass flux of phase n (2.2)
k_n	: average fluctuational kinetic energy of phase n (3.5)
M_n	: exchange of mass (3.2)
m_p	: mass of a particle
n_p	: number density of the particles
$\mathbf{n}_n$	: unit vector normal to the interface, pointing outwards phase n ($\mathbf{n}_p$ is sometimes written $\mathbf{n}$) (3.1)
$\mathbf{P}_n$	: exchange of momentum (3.3)
$P_{osm}(\phi)$	: osmotic pressure (1.3)
$\mathbf{Q}$	: total heat flux (3.5)
S_n	: exchange of entropy (3.6)
$\mathbf{u}$	: volume-weighted average velocity of the suspension (1.1)
$\mathbf{v}'_n$	: fluctuation relative to $\mathbf{v}_n$ (3.2)

$\mathbf{v}_I$	:	velocity of interfaces (3.1)
$\mathbf{w}$	:	relative velocity between particles and fluid (2.2)
$\mathbf{w}_n$	:	velocity of phase n relative to $\mathbf{v}$ (3.2)
χ_n	:	function of presence for phase n (3.1)
δ_I	:	function of presence of the interfaces (3.1)
Δ_I°, Δ_I	:	entropy production at interfaces (3.6)
ϕ_n	:	volume fraction (ϕ stands for ϕ_p) (3.1)
ϕ_M	:	volume fraction at maximum packing (1.5)
Φ	:	mechanical dissipation rate per unit volume (4.4)
η_n	:	shear viscosity (1.2)
$\eta(\phi)$	:	effective viscosity ratio function (1.2)
λ_n	:	thermal conductivity (1.6)
μ_n°, μ_n, μ^*, μ	:	chemical potentials per unit mass (2.2) and (2.3)
Π	:	total momentum flux (3.3)

Convective time-derivatives

$d/dt \equiv \partial/\partial t + \mathbf{v}.\nabla$

$d_n/dt \equiv \partial/\partial t + \mathbf{v}_n.\nabla$

1. THE FLUID-MECHANICAL APPROACH

The micro-structure of a multi-phase mixture is quite different from that of a molecular mixture. In a multi-phase mixture, the smallest parts containing a single chemical species have a size much larger than the molecular size. In other words, a multi-phase mixture appears as a three-dimensional patchwork made from rather large single-phase pieces. Hence, air is a molecular mixture, while smoke or fog are multi-phase mixtures, and brine is a molecular mixture while milk or paint are multi-phase mixtures. In fact, it is difficult to define a "critical size" for the basic pieces and the transformation from fog to wet air can be performed in a continous way . Consequently, one will not be surprised if multi-phase mixtures sometimes behave like molecular mixtures. However, the existence of a patch-work of continuous media will be essential for all what follows.

1.1 *The conservation law*

Modelling the behaviour of multi-phase mixtures is not an easy task and that difficulty is the very reason why so many different model equations have been proposed to describe them. Yet, there exists a minimum set of equations on which everyone agrees, those expressing the conservation laws for mass, momentum and energy. For all multi-phase mixtures (and for all molecular mixtures too), with a mass per unit volume $\rho(x,t)$, a velocity $\mathbf{v}(x,t)$ and a total energy per unit mass $e(x,t)$, these three conservation laws have the same well-known expressions, provided the external forces act with equal intensity on the unit mass of each component (as is the case for the gravitational force $\mathbf{g}$). These standard expressions are

$$\partial\rho/\partial t + \nabla.\, \rho\mathbf{v} = 0 \qquad (1.1.1)$$

$$\rho\, d\mathbf{v}/dt = \nabla.\Pi + \rho\mathbf{g} \qquad (1.1.2)$$

$$\rho\, de/dt = \nabla.(\mathbf{v}.\Pi - \mathbf{Q}) + \rho\mathbf{g}.\mathbf{v} \qquad (1.1.3)$$

where Π is the stress tensor and $\mathbf{Q}$ is the heat flux of the whole mixture, while $d/dt \equiv \partial/\partial t + \mathbf{v}.\nabla$ is the material time-derivative. In a pure fluid, Π and $\mathbf{Q}$ have the well-known expressions derived by Newton and Fourrier. In a multi-component mixture, these fluxes have more complicated expressions and a few examples will be given in sections 2 for Π and 6 for $\mathbf{Q}$. Besides the three above equations, there is still one more type of equation on which everybody agrees and it concerns the conservation of the mass of a component in the mixture which is written as

$$\partial\rho_n/\partial t + \nabla.\, \rho_n\mathbf{v}_n = 0 \qquad (1.1.4)$$

where ρ_n is the mass of component n per unit volume of the mixture, and $\mathbf{v}_n$ is its average velocity. When chemical reactions or phase changes occur between the different components, a source term appears in the right-hand side of (1.1.4), but we will delete that possibility in the first two chapters. Since the mass of the mixture is the sum of the masses of all components, the compatibility between the N equations (1.1.4) and equation (1.1.1) requires

$$\rho = \sum_n \rho_n \tag{1.1.5}$$

and

$$\rho \mathbf{v} = \sum_n \rho_n \mathbf{v}_n \tag{1.1.6}$$

Besides ρ_n, another important quantity is the volume occupied by component n per unit volume of the mixture. Let ϕ_n be that volume fraction. Since the whole volume is shared by the N components,

$$\sum_n \phi_n = 1 \tag{1.1.7}$$

If component n is made of an incompressible material, this means that $\rho_n{}^\circ$, the mass per unit volume of pure component n, is a constant. Then, the proportionality between the mass and the volume expresses as

$$\rho_n = \phi_n \rho_n{}^\circ \tag{1.1.8}$$

This allows to transform the mass conservation (1.1.4) into an equation for volume conservation

$$\partial\phi_n/\partial t + \nabla . \phi_n \mathbf{v}_n = 0 \tag{1.1.9}$$

Note here that the volume fraction is a quantity that cannot be defined in a molecular mixture. A conservation law like (I.1.9) is specific of multi-phase mixtures, while (I.1.4) holds for molecular mixtures too. If all components can be considered as incompressible, (1.1.7) and (1.1.9) imply

$$\nabla . \mathbf{u} = 0 \tag{1.1.10}$$

where

$$\mathbf{u} = \sum_n \phi_n \mathbf{v}_n \tag{1.1.11}$$

is the volumetric flux of the whole mixture. It is noteworthy thet **u** is a volume-weighted average velocity which is generally different from the velocity **v** which is mass-weighted (cf.(1.1.6)). In a mixture of incompressible components, it is **u** (and not **v**) which has a vanishing divergence since, as evident from (1.1.5) and (1.1.8), a constant value for any $\rho_n{}^\circ$ does not mean a constant ρ.

Whether component n is incompressible or not, a new velocity $\mathbf{v}_n$ appears in the mass or volume conservation. The problem of finding an explicit expression or equation for $\mathbf{v}_n$ is one more modelling problem (besides that concerning Π and $\mathbf{Q}$) and it will be dealt with in sections 3 to 5 for some peculiar types of mixtures.

1.2 *Modelling of the stress tensor*

In the continuum description of a pure fluid, a point is understood as a small volume containing many molecules, and quantities involved in the continuum description are meant as averaged values over that tiny volume. Similarly, in the continuum description of a multi-phase mixture, a point is to be figured as a small volume containing many pieces of each phase, and macroscopic quantities (such as $\rho(x,t)$ or $\Pi(x,t)$) must be understood as averaged values over such an elementary volume surrounding point $\mathbf{x}$ at time t. As a consequence, Π will appear as a sum of contributions from each phase, each contribution being proportional to the volume fraction of that phase. In case of a fluid(f)-particle(p) suspension one is led to write

$$\Pi_{ij} = \phi < \sigma_p{}^\circ >_{ij} + (1-\phi) < \sigma_f{}^\circ >_{ij}$$

where the brackets represent some averaging procedure (to be precised in chapter 3) while the superscript $^\circ$ stands for a "local" value which is usually fluctuating strongly in space and time. The local stress in phase n is represented by $\sigma_n{}^\circ$. The actual expression of Π also involves contributions from velocity fluctuations in both phases as well as contributions from the average relative motion between the particles and the fluid (see eq.(3.3.14)) but we here limit the arguments at some intuitive level. When the particles are incompressible and dispersed in a incompressible fluid, one usually introduces an isotropic pressure stress tensor and writes

$$\Pi_{ij} = -p\,\delta_{ij} + \phi < \tau_p{}^\circ >_{ij} + (1-\phi) < \tau_f{}^\circ >_{ij} \qquad (1.2.1)$$

where $\tau_f{}^\circ$ is the fluid viscous stress while $\tau_p{}^\circ$ is the traceless part of the particle stress $\sigma_p{}^\circ$. For an incompressible and newtonian fluid, $\tau_f{}^\circ = 2\eta_f e_f{}^\circ$ where η_f is the fluid viscosity and $e_f{}^\circ$ is the local deformation rate of the fluid, represented by a trace-less tensor. As a consequence, $< \tau_f{}^\circ >_{ij} = 2\eta_f < e_f{}^\circ >_{ij}$. Similarly, one introduces a local deformation rate $e_p{}^\circ$ everywhere inside a particle. For a particle with a constant volume, $e_p{}^\circ$ is a traceless tensor. It must be stressed that $<e_p{}^\circ>$ is not linked to the gradient of $\mathbf{v}_p$. In fact, $e_p{}^\circ$ and $< e_p{}^\circ >$ vanish for rigid particles, while nothing prevents $\nabla \mathbf{v}_p$ to be non-zero if the average velocity $\mathbf{v}_p$ changes from one point of the mixture to another. In short, $e_p{}^\circ$ and $e_f{}^\circ$ represent the small scale deformation rates of the two components. The average deformation rate for the whole mixture is defined as

$$e_{ij} = \phi < e_p{}^\circ >_{ij} + (1-\phi) < e_f{}^\circ >_{ij}$$

Since the local velocity is continuous (due to the boundary condition at the interfaces) it can be shown that the average deformation rate is related to the gradient of the average velocity $\mathbf{u}$ (and not the $\mathbf{v}$ gradient)

$$e_{ij} = (1/2)\,(\,\partial u_i/\partial x_j + \partial u_j/\partial x_i\,) \qquad (1.2.2)$$

Due to the assumptions concerning the incompressibility of the particles and the fluid, the trace e_{kk} vanishes (cf. (1.1.10)). Gathering the above results one deduces

$$\Pi_{ij} = -p\,\delta_{ij} + 2\eta_f e_{ij} + \phi < (\tau_p{}^\circ - 2\eta_f e_p{}^\circ)_{ij} > .$$

For a homogeneous suspension, the above result is equivalent to [1]

$$\Pi_{ij} = -p\,\delta_{ij} + 2\eta_f e_{ij} + n_p << \int (\tau_p{}^\circ - 2\eta_f e_p{}^\circ)_{ij}\, dV >> \qquad (1.2.3)$$

where $n_p(x,t)$ is the average number of particles per unit volume of the mixture and the integral is over the volume of a test-particle with center fixed at the point where the average is performed. The value of the integral will depend on the various possible orientations and shapes of the test-particle, as well as on the spatial distribution of all other particles in the suspension, and the double brackets stand for an average over all those possibilities. The problem of finding Π is thus reduced to the problem of calculating the relative stress $\tau_p{}^\circ - 2\eta_f e_p{}^\circ$ all over the test-particle for a given configuration, and then to determine its average value with the configuration probability. Note that this relative sress emphasizes the difference of mechanical behaviour between the particle and the surrounding fluid. If the particle was made from a fluid material with the same viscosity ($\eta_p = \eta_f$), the relative stress would vanish and, as expected, the mixture would behave as a pure fluid of viscosity η_f. Conversely, if the particle is a solid or a fluid with a different viscosity, the last term in (1.2.3) will not vanish. The integral over the volume of the test-particle can be rewritten in a form more favourable for calculations [2]. From the definition of $e_p{}^\circ$ and the continuity of velocity at the interfaces one gets

$$\int (e_p{}^\circ)_{ij}\, dV = (1/2) \int (\mathbf{v}_f{}^\circ \otimes \mathbf{n} + \mathbf{n} \otimes \mathbf{v}_f{}^\circ)_{ij}\, dS \qquad (1.2.4)$$

where $\mathbf{n}$ is the unit normal pointing outwards the particles. Somewhat analoguously, the continuity of forces at the interfaces (upon neglect of surface tension) leads to

$$\int (\sigma_p{}^\circ)_{ij}\, dV = \int r_j (\sigma_f{}^\circ.\mathbf{n})_i\, dS - \int r_j (\nabla.\sigma_p{}^\circ)_i\, dV$$

where $\mathbf{r}$ is the position relative to the sphere center. It is then easy to deduce $\int \tau_p{}^\circ\, dV$ from the symmetric and traceless part of the r.h.s. of the above result. In case of particles with negligible inertia, $\nabla.\sigma_p{}^\circ$ equals the external force acting per unit volume of the particles, and if that external force is gravity the last integral vanishes. The nice feature of particles with negligible inertia is thus the possibility of computing $\int \tau_p{}^\circ\, dV$ from the symmetric and traceless moment of the fluid force $\sigma_f{}^\circ.\mathbf{n}$ over the particle surface

$$\int (\tau_p{}^\circ)_{ij}\, dV = (1/2) \int [r_j (\sigma_f{}^\circ.\mathbf{n})_i + r_i (\sigma_f{}^\circ.\mathbf{n})_j - (2/3)(\mathbf{r}.\sigma_f{}^\circ.\mathbf{n})\,\delta_{ij}]\, dS \quad . \qquad (1.2.5)$$

Inserting (1.2.4) and (1.2.5) into (1 2 3) leads to Batchelor's expression for the suspension stress. On general grounds, the fluid force $\sigma_f^\circ.\mathbf{n}$ will depend on the type of flow in which the particle is embedded, on the position and orientation of the particle in that flow, as well as on the configuration of the neighbouring particles. To perform the average << >> on the test-particle, one needs to know the probability density for all possible orientations of the test-particle, and for all possible configurations of its neighbours, and this is the most difficult part of the job. A few examples are provided below.

SUSPENSIONS OF PLATES : To begin with, let us consider an instructive (although somewhat artificial) example. Two parralel surfaces are separated by a fixed gap h (cf.figure below). A fluid is introduced in the gap and the upper surface is moved with a constant velocity V. One measures the force necessary to obtain a stationary motion. Per unit surface area, that force f_1 equals the stress Π_{12} exerted by the fluid on the upper plate. For a newtonian fluid with viscosity η_f, $\Pi_{12} = \eta_f\, \partial v_1/\partial x_2$. The balance of forces in a slice of fluid requires the velocity gradient to be a constant all over the gap, hence $f_1 = \eta_f V/h$. Repeat now the experiment with a multi-phase mixture made of the above newtonian fluid and a collection of rigid plates (with $\rho_p^\circ = \rho_f^\circ$ to avoid settling), all of them parralel to the two surfaces bounding the flow The sum of the thick nesses of the plates is a fraction ϕ of the total gap h. The balance of forces in the fluid part of the mixture still requires a constant value for the fluid velocity gradient. But this gradient is now larger by a factor (1-ϕ) because of the presence of the plates. One then deduces $f_1 = (\eta_f/(1-\phi))\, V/h$ or equivalently

$$\Pi_{12} = (\eta_f/(1-\phi))\, e_{12}\,.$$

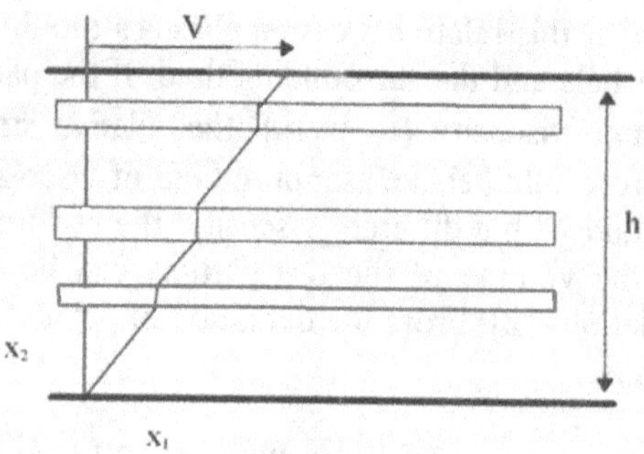

Fig. 1 : a suspension of parralel plates in a shear flow

Everything happens as if the suspension of plates behaves as a fluid of effective viscosity $\eta_f/(1-\phi)$. Note that this result concerns a very special position of the plates inside the gap and one component of the stress tensor only Note also that the effective viscosity is always larger than that of the pure fluid, and that it diverges when the plates occupy the whole available place. These results are in fact displayed qualitatively by any mixture.

SUSPENSIONS OF RIGID SPHERES : A rigid sphere (radius a) is introduced in a flow with homogeneous velocity gradients. The solution of Stokes equation with due boundary conditions on the sphere surface allows to compute the r.h.s. of (1.2.5). A rigid sphere eliminates all problems concerned with the orientation and shape of the test-particle. The << >> average will not modify the single-particle result provided the suspension is dilute enough, because the configurations with one particle (or more) close to the test-particle will have a vanishingly small probability. One then finds

$$\Pi_{ij} = -p\,\delta_{ij} + 2\eta_f(1+5\phi/2)\,e_{ij} \qquad (1.2.6)$$

where $\phi = n_p\,(4\pi a^3/3)$ is the volume fraction of the particles, of order 10^{-3} or less for result (1.2.6) to hold. Hence, the mechanical behaviour of a dilute suspension of spheres is that of a newtonian fluid with effective viscosity $\eta_f\,(1+5\phi/2)$, a result obtained by Einstein at the beginning of that century . Note that this effective viscosity holds whatever the type of flow (shear or elongational) represented by e_{ij}. For more concentrated suspensions, the elementary force $\sigma_f^\circ.\mathbf{n}$ will depend on the spatial configuration of the neighbouring particles and the double brackets average appearing in (1.2.3) will depend on the probability of occurence of a given configuration, which itself depends on the type of flow considered.. The corresponding calculations are extremely difficult and do not extend much the range of validity of Einstein's result. For that reason, one cannot but resort to phenomenology, and propose more or less happy guesses to generalize (1.2.6). Since the concentration of particles is certainly an important parameter for the effective viscosity, a widely used phenomenological expression is

$$\Pi_{ij} = -p\,\delta_{ij} + 2\eta_f\,\eta(\phi)\,e_{ij} \qquad (1.2.7)$$

where $\eta(\phi)$ is the effective reduced viscosity of the suspension. This effective viscosity is getting tremendous values for some maximum volume fraction ϕ_M above which velocity gradients are extremely small and the suspension behaves as a solid phase. Many expressions for $\eta(\phi)$ have been proposed heretofore and examples may be found in the present volume in U. Schaflinger's and M. Ungarish's lectures. Such expressions consider that the suspension of spheres behaves as a newtonian fluid whatever is the particle concentration. It can provide a good modelling for some suspensions but it must be kept in mind that it relies on two assumptions which may fail for others: it supposes that the probability of occurence of a given configuration of particles is independent of the type of flow and that it is not influenced by the Brownian motion of particles.

SUSPENSION OF RIGID FIBERS : These particles are elongated cylinders of length L and diameter d. The quantity $r = L/d$ is known as the aspect ratio and the following results will concern particles with a very <u>large aspect ratio</u> only. The elementary force $\sigma_f^\circ.\mathbf{n}$ acting on the surface of a rigid fiber (introduced in a fluid flow with homogeneous velocity gradients) is computed as a function of the particle orientation vector $\mathbf{p} = \mathbf{L}/L$. The force and the torque acting on the long fiber are then determined and for a neutrally-buoyant fiber ($\rho_p^\circ = \rho_f^\circ$) with no external torque acting on it, the following conclusions hold: the fiber translates with the fluid and it rotates with an angular velocity which equals the fluid rotating velocity plus a correcting term depending on its orientation

$$\mathbf{\Omega}^p = \mathbf{\Omega} + \mathbf{p} \times (\mathbf{e}.\mathbf{p}) \qquad (1.2.8)$$

Then, with the help of (1.2.5), one deduces the stress tensor of a dilute suspension of fibers from (1.2.3). The result is

$$\Pi_{ij} = - p\,\delta_{ij} + 2\,\eta_f e_{ij} + 2\,\eta_f\,\phi(2r^2/3\mathrm{Log}(2r))\,[\langle\langle p_i p_j p_k p_l \rangle\rangle - \langle\langle p_k p_l \rangle\rangle\,\delta_{ij}/3]\,e_{kl} \quad (1.2.9)$$

where $\phi = n_p\,\pi d^2 L/4$ is the volume fraction of the rods. That expression, obtained from a result concerning a single particle, is restricted to suspensions that are dilute enough for the rotating motion of one rod not to interfere with the rotating motion of another rod. This means that each particle must be free to occupy a volume much larger than L^3 so that the volume fraction must be much lower than r^{-2}, a quite small value. Diluteness also implies that the << >> average involves the probability distribution for orientation, that probability depending itself on the type and intensity of the flow as well as on the intensity of brownian motion. It happens that for higher concentrations (limited to $\phi < r^{-1}$ however), the stress tensor has a rather similar expression but with $2r^2/3\mathrm{Log}(2r)$ replaced by $f(r,\phi) = 4r^2/3\mathrm{Log}(\pi/\phi)$ [3]. In fact, this extension is rigorous for elongational flows only but, for simplicity, we here suppose it to hold whatever e_{ij} is. The concentration range between r^{-2} and r^{-1} is called "semi-dilute". In a semi-dilute suspension of rods, the particles volume fraction is still much less than one but it is remarkable that the product $\phi f(r,\phi)$ can be very large. One thus expects large stresses and also stresses that depend on the type of flow because of its influence on the average orientation of the rods (for more details see the lectures by G. Cognet). For example, in a shear flow directed along direction 1 and varying along direction 2, all components of e_{ij} vanish except $e_{12} = e_{21}$. The effective shear viscosity relative to the shear viscosity of pure fluid is then

$$\eta_{shear} = \Pi_{12} / 2\eta_f e_{12} = 1 + 2\,\phi\,f(r,\phi)\,\langle\langle p_1^2 p_2^2 \rangle\rangle \quad .$$

If instead of a shear flow, one considers an elongational flow with a symmetry axis along direction 1, the only non-vanishing components are $e_{22} = e_{33} = -e_{11}/2$ and the effective elongational viscosity is defined as

$$\eta_{elong} = [\Pi_{11} - (\Pi_{22}+\Pi_{33})/2] / 3\eta_f e_{11} = 1 + (\phi f(r,\phi)/6)\,[9\langle\langle p_1^4 \rangle\rangle - 6\langle\langle p_1^2 \rangle\rangle + 1]$$

Experimental observations reveal that the fibers are mostly aligned with the velocity field, be it a shear or an elongational flow. This means that $\langle\langle p_1^2 p_2^2 \rangle\rangle$ is very small in the shear flow while $\langle\langle p_1^4 \rangle\rangle \cong \langle\langle p_1^2 \rangle\rangle \cong 1$ in the elongational flow, so that elongational flows will display a considerable effective viscosity as compared to that in shear flows.

The action of the fluid flow on the particle orientation can be described more precisely. As already mentionned, in the absence of external torques, a neutrally-buoyant fiber moves with the fluid and rotates with an angular velocity given in (1.2.8). In a continuum description of a suspension of fibers, one introduces a vector-field $\mathbf{p}(x,t)$ and transforms the single particle result into

$$d\mathbf{p}/dt = (\mathbf{p}.\nabla)\,\mathbf{u} - (\mathbf{p}.\mathbf{e}.\mathbf{p})\,\mathbf{p} \qquad (1.2.10)$$

One can check that this equation guarantees that **p.p** is a constant. Expression (1.2.9) for the stress tensor is bound to the average values $<<p_ip_j>>$ and $<<p_ip_jp_kp_l>>$. From (1.2.10), one can deduce the evolution of p_ip_j and ,after averaging, it becomes.

$$d <<p_ip_j>> /dt = <<p_ip_k>> \partial u_j/\partial x_k + <<p_jp_k>> \partial u_i/\partial x_k - 2 <<p_ip_jp_kp_l>> e_{kl} \quad .(1.2.11)$$

Two important points are to be stressed here:

a) To solve for $<<p_ip_j>>$ in a given velocity field, one needs to know $<<p_ip_jp_kp_l>>$. That fourth moment of the orientation distribution obeys itself an evolution equation (also deduced from (1.2.10)) where appear still higher moments. That sequence of interwoven equations cannot be solved unless it is truncated at some level. The lowest level expresses $<<p_ip_jp_kp_l>>$ in terms of $<<p_ip_j>>$. Then, the suspension stress can be completely computed from the solution of (1.2.11). The compatibility with experimental resuls is the ultimate test to justify the assumed relation between the moments of the orientation distribution.

b) According to (1.2.10), the vector $\mathbf{L} = L\mathbf{p}$ evolves as

$$d\mathbf{L}/dt = (\mathbf{L}.\nabla)\, \mathbf{u} - (\mathbf{p}.\mathbf{e}.\mathbf{p})\, \mathbf{L}$$

while if **L** was to represent a small <u>fluid filament</u> in a flow with homogeneous velocity gradients, its evolution in time would be

$$d\mathbf{L}/dt = (\mathbf{L}.\nabla)\, \mathbf{u} \qquad (1.2.12)$$

This means that the quantity $d\mathbf{p}/dt - (\mathbf{p}.\nabla)\mathbf{u}$ is a good candidate to witness to the difference of mechanical behaviour between the rigid fibers and the fluid. The link with the general result (1.2.3) is clearly displayed if one rewrites the last term of (1.2.9) as

$$n_p << \eta_f V^{eff} (p_ip_k\, \partial u_j/\partial x_k + p_jp_k\, \partial u_i/\partial x_k - dp_ip_j /dt\,) >> \qquad (1.2.13)$$

where $V^{eff} = \pi L^3/6Log(2r)$ is some effective volume of the rigid fiber, much larger than the true volume for fibers with large aspect ratio.

1.3. *Modelling of the relative motion*

Let us begin by the motion of a single particle in a unbounded and uniformly moving fluid. The fluid is responsible for two forces acting on the particle, the Archimede's thrust and a viscous force which opposes the relative motion. A small spherical particle (radius a and mass m_p) will often have a negligible inertia and the balance of forces will appear as

$$0 \cong 6\, \pi a\, \eta_f (\mathbf{V} - \mathbf{V}^p) - (4\, \pi a^3/3)\, \rho_f^{\circ}\, \mathbf{g} + m_p\, \mathbf{g} \quad . \qquad (1.3.1)$$

The above equation can be extended without difficulty to a very dilute suspension containing n_p spheres per unit volume

$$0 \cong 6\pi\, n_p\, a\, \eta_f\,(\mathbf{v}_f - \mathbf{v}_p) + n_p m^* \mathbf{g}$$

where $\mathbf{v}_p$ and $\mathbf{v}_f$ are the average values of the particles and fluid velocities while $m^* = (4\pi a^3/3)(\rho_p^\circ - \rho_f^\circ)$ is the apparent mass of a particle. Such a momentum balance implies that in a motionless fluid, all the particles will be driven in the direction of $m^*\mathbf{g}$ and that the process will stop at some bottom wall for particles with a positive m^*. This conclusion happens to fail for tiny particles which at the end of the sedimenting process are distributed in space with a concentration profile given by $n_p = n_{po} \exp(-\, m^*\mathbf{g}.\mathbf{x}/k_B T)$ where k_B is the Boltzmann constant and T is the temperature of the suspension. That profile can be accounted for by introducing a new force acting on the particles, proportional to the gradient of the particle concentration, and which transforms the force balance into

$$0 \cong 6\pi\, n_p\, a\, \eta_f\,(\mathbf{v}_f - \mathbf{v}_p) + n_p m^* \mathbf{g} - k_B T\, \nabla n_p \quad . \qquad (1.3.2)$$

The importance of the particle size is easy to understand if one remembers that m^* is of order a^3. For large particles, the apparent mass will overcome the gradient of concentration force and the sedimentation process will be complete, while smaller particles will display a concentration profile. A "critical" size is not easy to define, but if the particles are introduced in a container of height L, then the higher the ratio $k_B T/m^* g L$, the more homogeneous is the equilibrium concentration.

The above force balance was written for dilute suspensions. Its extension to non-dilute ones is a challenge and, as was the case for the stress tensor, one is more or less obliged to resort to phenomenology. A widely used extension of (1.3.2) is

$$0 \cong \phi(\rho_p^\circ - \rho_f^\circ)\mathbf{g} + (9\eta_f/2a^2)\,\phi(1-\phi)\, h^{-1}(\phi)\,(\mathbf{v}_f - \mathbf{v}_p) - \nabla P_{osm}(\phi) \qquad (1.3.3)$$

where $h(\phi)$ and $P_{osm}(\phi)$ are two functions of the particle volume fraction with the following physical interpretation: $h(\phi)$ describes the hindered settling of the particles when ϕ increases while $P_{osm}(\phi)$ is the osmotic pressure of the particles (cf. section 2.3) which characterizes the change of intensity of the concentration gradient force. Many expressions have been proposed for these two functions. Concerning P_{osm}, see (2.3.16) for hard-spheres particles. Concerning h, the well-known Richardson-Zaki expression is $h = (1-\phi)^n$ with $n \cong 5$. The above balance of forces for a non-dilute suspension can also be presented as

$$\mathbf{v}_p = \mathbf{u} + h(\phi)\, \mathbf{v}_{sed} - D\, \nabla\phi/\phi \quad . \qquad (1.3.4)$$

where $\mathbf{v}_{sed} = (2a^2/9\eta_f)\,(\rho_p^\circ - \rho_f^\circ)\mathbf{g}$ is the sedimentation velocity of a single particle in a unbounded fluid, while D is the coefficient of diffusion of the particles in the mixture. Its dilute limit value is $D_o = k_B T/6\pi\eta_f a$ (Einstein again !) while for more concentrated suspensions

$$D = (2a^2/9\eta_f)\, h(\phi)\, \partial P_{osm}/\partial\phi \qquad . \qquad (1.3.5)$$

The role of the concentration gradient in the relative flow of particles is quite similar to that found in molecular mixtures (Fick's law) but the role of external forces displayed by $\mathbf{v}_{sed}$ in (1.3.4) is important for multi-phase mixtures only.

Result (1.3.4) achieves our goal which was to find the average velocity $\mathbf{v}_p$ that was introduced in the mass conservation law. It should be clear that to obtain (1.3.4), we supposed the fluid motion to be gentle enough. For highly accellerated fluid flows, the forces exerted by the fluid will be different and the inertia of the particles will come into play. In particular, the Archimede's thrust is transformed into $\phi\rho_f^\circ(d_f\mathbf{v}_f/dt - \mathbf{g})$ where $d_f/dt \equiv \partial/\partial t + \mathbf{v}_f.\nabla$, while the interphase friction force will be written as $-\,R(\mathbf{v}_p-\mathbf{v}_f)$ where R is some friction coefficient. Taking the particle inertia into account, as well as a possible force $\mathbf{F}_c$ due to contacts or collisions between particles, the momentum balance of the particles will be written as

$$\phi\rho_p^\circ\, d_p\mathbf{v}_p/dt = \phi\rho_f^\circ\, d_f\mathbf{v}_f/dt + \phi(\rho_p^\circ - \rho_f^\circ)\mathbf{g} - R(\mathbf{v}_p-\mathbf{v}_f) + \mathbf{F}_c - \nabla P_{osm}(\phi) \qquad . \qquad (1.3.6)$$

Concerning the fluid phase, its momentum balance can be deduced fom (1.3.6) and (1.1.2) as

$$\rho_f^\circ\, d_f\mathbf{v}_f/dt = \rho_f^\circ\mathbf{g} + \nabla.\sigma + R(\mathbf{v}_p-\mathbf{v}_f) - \mathbf{F}_c + \nabla P_{osm}(\phi) \qquad (1.3.7)$$

where the stress σ differs from Π by a kinetic contribution related to the relative motion

$$\sigma = \Pi + (\rho_p\rho_f/\rho)\,(\mathbf{v}_p-\mathbf{v}_f)\otimes(\mathbf{v}_p-\mathbf{v}_f) \quad .$$

Comparing the particles momentum balance (1.3.6) with its simplified form (1.3.3), one sees the importance of deciding if inertia forces will be important or not, if the osmotic or contact force can be neglected or not. There are no general recipies and one must decide according to the type of suspension and the type of flow considered. We will now examine two particular cases.

1.4 *Flow through porous media*

The flow of a fluid through porous rocks is a particular case of a relative flow between fluid and particles. One has just to imagine the porous rocks as a suspension of particles with a high enough concentration for the particles to be in permanent contact with each other. Actually, this picture is relevant to what is called unconsolidated porous media, but the consolidated ones for which the particles are somehow "glued" to each other are not fundamentally different as far as the relative flow is concerned.

Let us first consider some particle pertaining to the porous rock. It is submitted to a force exerted by the fluid on its surface and to a force exerted by the neighbouring particles through contacts. Then, if we consider all the particles contained in a small representative volume of the rock, the fluid force will be distributed all over that volume, while the contact force will concern the only particles lying on the boundary of the representative volume. The reason is that all contact forces between particles inside that representative volume will cancel each other. Hence, when written per unit volume of the mixture (fluid+rocks), the contact force will appear as the divergence of some stress tensor σ_c

$$\mathbf{F}_c = \nabla.\sigma_c$$

If the relative flow is slow enough, the interphase friction force (previously written as $R(\mathbf{v}_p-\mathbf{v}_f)$ for suspensions) will be written as $(\eta_f/K)(\mathbf{v}_p-\mathbf{u})$ where K is the permeability of the porous medium and $\mathbf{u}$ the volume flux defined in (1.1.11). Concerning the total stress tensor of the mixture, it will include an isotropic pressure stress, the contact stress, and some possible viscous stress. Generally, the viscous stress is negligible when compared to the contact stress and one writes simply

$$\sigma = -\, p\, I + \sigma_c\ .$$

The osmotic force is of no concern for a porous medium and the above modelling for $\mathbf{F}_c$ and σ transform the general equations (1.3.6) and (1.3.7) into

$$\phi\rho_p{}^\circ\, d_p\mathbf{v}_p/dt = \phi\rho_f{}^\circ\, d_f\mathbf{v}_f/dt + \phi(\rho_p{}^\circ - \rho_f{}^\circ)\mathbf{g} - (\eta_f/K)(\mathbf{v}_p-\mathbf{u}) + \nabla.\sigma_c \qquad (1.4.1)$$

and

$$\rho_f{}^\circ\, d_f\mathbf{v}_f/dt = \rho_f{}^\circ\mathbf{g} - \nabla p + (\eta_f/K)(\mathbf{v}_p-\mathbf{u})\ . \qquad (1.4.2)$$

As expected, the contact stresses play a role in the particulate phase only. Associated with the two mass equations (1.1.4) and explicit expressions for K and σ_c (which depend on the particle concentration mainly), these are the basic equations to study elastic porous media. They are also used to study the compaction of sediments, foams or any other problem in which the permanent contact between particles is important.

A very extreme case is a porous rock so rigid that any motion and deformation of the porous matrix can be neglected. In other words, small deformations induce so large contact stresses that the solution of the momentum equation (1.4.1) and the volume conservation (1.1.9) can be expressed with good accuracy by

$$\mathbf{v}_p = 0 \qquad \text{and} \qquad \partial\phi/\partial t = 0\ .$$

If the fluid is incompressible and the fluid inertia can be neglected, then the mass and momentum equations for the fluid phase are reduced to

$$\nabla.(1-\phi)\,\mathbf{v}_f = 0 \qquad (1.4.3)$$

and

$$(1-\phi)\,\mathbf{v}_f = (K/\eta_f)\,(\rho_f{}^\circ\mathbf{g} - \nabla p)\ . \qquad (1.4.4)$$

These are the famous Darcy's equations describing the slow motion of an incompressible fluid through a rigid porous medium. The velocity $(1-\phi)\mathbf{v}_f$ is sometimes called the filtration velocity because the fluid flux through a macroscopic surface element $\mathbf{n}$ dS of the porous medium is $(1-\phi)\mathbf{v}_f \cdot \mathbf{n}dS$, due to the fraction $1-\phi$ of the surface dS which is available to the fluid flow. When the fluid inertia can no longer be neglected, (1.4.4) is replaced by (1.4.2) where K possibly depends on the Reynolds number $\rho_f^\circ v_f a/\eta_f$ (a represents an order of magnitude of the pore size).

1.5. *Fluidized beds*

A fluidized bed refers to a suspension in which an upwards flow of fluid prevents heavier particles to sediment. In a fluidized bed furnace for example, pulverized coal is suspended by a rapid air flow, and the large amount of interfaces in contact with oxygen results in a large increase of the combustion rate as compared to more classical furnaces. To work correctly, that device needs a flow rate which is neither too slow (for the coal particles not to soot the furnace) neither too rapid (for the particles not to escape before complete burning). What is the range of flow rates needed for a good working order ?

Starting from a situation where the heavy particles are stacking on a bottom grid, an increasing fluid flow rate is forced to pass through the grid. At first, the particles behave as a compact sediment analoguous to some unconsolidated porous medium and the pressure difference between the top and the bottom is proportional to the flow rate (cf.(1.4.4)), while the intensity of the contact stresses is progressively decreasing. Then, above a minimum flow rate Q_{min}, the compact sediment seems to puff up, the particles are no longer in permanent contact with each other and the contact stresses boil down to collisions stresses. The former compact sediment is now transformed into a suspension which is all the more dilute that the fluid flow rate is higher. Finally, when the flow rate exceeds a maximum value Q_{max}, all particles are blowed up. The fluidized bed regime with motion-less particles needs a flow rate between Q_{min} and Q_{max}.

The simplest modelling of fluidized beds neglects viscous shear stresses (because the average velocity of the mixture is quasi-uniform over the cross-section of the device). Except for the top part of the fluidized-bed, the concentration of particles and the relative motion are fairly constant so that the osmotic force and the collision force dissapear (the collision stress is a constant). Moreover, if we are interested in stationary situations only, the average particle velocity vanishes and the momentum equations (1.3.6) and (1.3.7) simplify to

$$0 \cong \phi(\rho_p^\circ - \rho_f^\circ)\mathbf{g} + R\mathbf{v}_f$$

and

$$0 \cong \rho_f^\circ \mathbf{g} - \nabla p - R\mathbf{v}_f$$

Whenever the friction coefficient can be given the Stokes-like expression proposed in (1.3.3), the above equations are equivalent to $\nabla p = \rho\ \mathbf{g}$ (where ρ is the mass per unit volume of the whole suspension), and to

$$(1-\phi)\ \mathbf{v}_f = -\ h(\phi)\ \mathbf{v}_{sed} \qquad (1.5.1)$$

Note that (1.5.1) is nothing but a particular case of (1.3.4) with $\mathbf{v}_p = 0$. If the cross-section S of the device is constant, then the fluid flow rate is $(1-\phi)S\ \mathbf{v}_f$. Since the hindrance factor $h(\phi)$ is a decreasing function of the particle volume fraction, it is clear from (1.5.1) that the fluidized bed can exist only in the range between $Q_{min} = S\ h(\phi_M)\ v_{sed}$ and $Q_{max} = S\ v_{sed}$, where $\phi_M \cong 0.6$ is some packing volume fraction.

1.6. *Modelling the heat flux*

The modelling of the suspension heat flux $\mathbf{Q}$ proceeds along the same steps as for the suspension stress Π. The "intuitive" expression of $\mathbf{Q}$ is a sum of contributions from each of the two phases and the analog of (I.2.1) is

$$\mathbf{Q} = \phi < \mathbf{q}_p{}^\circ > + (1-\phi) < \mathbf{q}_f{}^\circ > \quad . \qquad (1.6.1)$$

If the Fourier heat conduction law is applicable to the suspending fluid, one deduces

$$\mathbf{Q} = -\lambda_f \nabla T + n_p << \int (\mathbf{q}_p{}^\circ + \lambda_f \nabla T_p{}^\circ)\, dV >> \quad , \qquad (1.6.2)$$

where λ_f is the thermal conductivity of the suspending fluid, while T is the average suspension temperature. Note that ∇T is also the average of the local temperature gradients due to the continuity of temperature at the interfaces. The above result is the analog of (1.2.3) for the stress tensor and it emphasizes the difference of thermal behaviour between the particles and the embedding fluid. If the particle's material also obbeys Fourier law, the volume integral vanishes whenever $\lambda_p = \lambda_f$. When the two thermal conductivities are different, the knowledge of the local temperature gradient $\nabla T_p{}^\circ$ all over the test-particle will lead to the particle contribution, provided one knows the various probabilities distributions involved in << >>. For the simplest case of a dilute suspension of spheres, it was found by Maxwell that

$$\mathbf{Q} = -\lambda_f [1 + 3\phi\, (\lambda_p-\lambda_f)/(\lambda_p+2\lambda_f)]\, \nabla T \qquad (1.6.3)$$

However, Maxwell's result holds for a completely static situation. In a flowing suspension, the motion occuring at the particle scale will modify the local temperature gradient $\nabla T_p{}^\circ$ as well as the probability distributions involved in the averaging procedure. As a result, the thermal conductivity of the suspension will depend on the type and strength of the flow, not to mention the possible role of Brownian motion. Last but not least, the "intuitive" expression (1.6.1) must be completed by terms linked to temperature and velocity fluctuations occuring at the particle scale, as well as by terms linked to the average relative motion (cf 3.5.16). By the way, the same remark holds for expression (1.2.1) of the stress tensor, the complete expression of which will be found in (3.3.14). Concerning the role of velocity fluctuations in the suspension heat flux, one cannot but recommend to read references [4] and [5].

2. THE THERMODYNAMICAL APPROACH

In the previous chapter we insisted on the mechanics of suspensions and developped models mainly concerned with stresses and relative motion. Yet, we have seen that the relative motion of small particles is influenced by a diffusion force involving the gradient of particle concentration. One of the aim of this chapter will be to find the origin and general expression of that force. More generally, we will consider the constraints imposed by the second law of thermodynamics on the macroscopic modelling of suspensions. We begin by a review of molecular mixtures. That review will be presented so as to emphasize the increasing complexities resulting from the increase of the particle size, and it will help us to make the link between multi-phase mixtures, colloidal suspensions and molecular mixtures.

2.1. *The first and second law of thermodynamics*

In the first chapter, the conservation laws for mass, momentum and energy were presented and their expression for continuous media was written in (1.1.1) to (1.1.3). Eq.(1.1.3) is nothing but the first law of thermodynamics which states that the change of total energy of a system is due to the work developped by the forces exerted on that system, and to the other forms of energy (heat for instance) exchanged with the outside world. The total energy is made of the kinetic and internal energy. For a mixture in which each component has its own velocity, several different definitions of the kinetic energy are possible. In all what follows we associate the kinetic energy with the mass-weighted velocity **v** only (see its definition (1.1.6)), and we will write the total energy per unit mass as

$$e = \varepsilon + v^2/2 \qquad (2.1.1)$$

This defines the internal energy ε per unit mass of the mixture. A simple manipulation of the momentum and energy equations leads to its evolution equation

$$\rho\, d\varepsilon/dt = \mathbf{\Pi} : \nabla \mathbf{v} - \nabla . \mathbf{Q} \qquad (2.1.2)$$

The second law of thermodynamics (or entropy law) is a statement concerning the evolution of the entropy of any system. Entropy can be thought of as a measure of the "degree of desorder", and for an isolated system it cannot decrease. When the system is no longer isolated, its entropy change cannot be less than the entropy exchanged with the outside world. For all continuous media, if **H** represents the entropy flux and s is the entropy per unit mass, the second law is expressed as

$$\rho\, ds/dt + \nabla . \mathbf{H} \geq 0 \qquad (2.1.3)$$

The aim of the thermodynamic approach is to detail the incidence of the constraint (2.1.3) on the modelling of a suspension of particles. The clue is to find between ε, s and other state variables, a relationship that holds not only for equilibrium situations, but also for flows in which thermodynamic equilibrium is not achieved. The following sections will precise those relationships and the implications of the second law for various kinds of suspensions.

2.2. *Binary molecular mixtures*

A gazeous mixture of oxygen and nitrogen, a solution of sugar in water are examples of molecular mixtures. They may be thought of as suspensions in which the particles have shrunk to a molecular size. The static properties and dynamic behaviour of molecular mixtures is rather well-known, but we will review them in a way which emphasizes their link with suspensions of supramolecular or macroscopic particles. To obtain the relationship between internal energy and entropy in a molecular mixture, we will start from another thermodynamic quantity which is the free-enthalpy. For a pure fluid (made from a single molecular species) the free enthalpy per unit mass is also called its chemical potential. It is a function of its pressure p and temperature T, noted $\mu^\circ(p,T)$. Roughly speaking, that chemical potential represents the energy change of the fluid when adding a unit mass at constant pressure and temperature. If the pure fluid is made of N molecules of mass m, its free-enthalpy is $G = Nm\,\mu^\circ(p,T)$. When one mixes together N_1 molecules of species 1 with N_2 molecules of species 2, the free-enthalpy of the mixture can be written quite generally as

$$G = N_1m_1\,\mu_1^\circ(p,T) + N_2m_2\,\mu_2^\circ(p,T) + (N_1m_1+N_2m_2)\,\Delta g \qquad (2.2.1)$$

where

$$\Delta g = c_1\,\Delta g_1(c_1,p,T) + c_2\,\Delta g_2(c_2,p,T) \qquad (2.2.2)$$

Here $c_n = N_nm_n/(N_1m_1+N_2m_2)$ is the mass fraction of species n $(c_1+c_2=1)$. The first two terms on the right-hand side of (2.2.1) are a mere addition of the energies of each component (as if they were separated from each other) and Δg represents a measure of how much the two species feel better when mixed together. Δg is called the free-enthalpy of mixing. If $\Delta g>0$, the mixing operation has little chance to succeed because the free-enthalpy of the mixture is then higher than the sum of the free-enthalpies when the two components are completely segregated. This is the case of two unsoluble liquids such as oil and water. Conversely, the mutual solubility of two species requires $\Delta g<0$ and this is the case we consider henceforth. The binding energy represented by Δg is a function of pressure and temperature, but most importantly it is a function of the mass fractions. This is because the change of energy upon adding the unit mass of one component is likely to depend on the relative amount of components already present in the mixture. Except for its negative value and its vanishing when c_1 or c_2 vanishes, we will not insist on the explicit expression of Δg. Suffice it to say that its dependence on the concentration is responsible for important dynamical phenomena such as the concentration diffusion. Before considering these dynamical aspects, we deal first with the equilibrium properties of molecular mixtures. Several definitions must be given. From (2.2.1) one deduces the chemical potential of each of the two species in the mixture

$$\mu_n = m_n^{-1}\,\partial G/\partial N_n = \mu_n^\circ + \Delta g + (1-c_n)\,\partial\Delta g/\partial c_n \qquad (2.2.3)$$

The entropy s and volume $1/\rho$ per unit mass of the mixture are defined as

$$s = -(N_1m_1 + N_2m_2)^{-1}\,\partial G/\partial T = c_1s_1 + c_2s_2 \qquad (2.2.4)$$

and

$$1/\rho = (N_1m_1 + N_2m_2)^{-1}\,\partial G/\partial p = c_1/\rho_1^\circ + c_2/\rho_2^\circ \quad , \qquad (2.2.5)$$

with the following expressions for the specific entropy s_n and the specific volume $1/\rho_n^\circ$

$$s_n = -\,\partial(\mu_n^\circ + \Delta g_n)/\partial T \qquad \text{and} \qquad 1/\rho_n^\circ = \partial(\mu_n^\circ + \Delta g_n)/\partial p .$$

It is interesting to note that s_n depends on concentration if Δg_n depends on temperature, while $1/\rho_n^\circ$ depends on concentration if Δg_n depends on pressure. As to the internal energy of the mixture, it is defined as

$$\varepsilon = g + Ts - p/\rho \qquad (2.2.6)$$

where $g = c_1\mu_1^\circ + c_2\mu_2^\circ + \Delta g$ is the free-enthalpy per unit mass. It is easy to check that

$$g = c_1\mu_1 + c_2\mu_2 \quad , \qquad (2.2.7)$$

then,

$$dg = (\mu_2-\mu_1)\,dc_2 - s\,dT + (1/\rho)\,dp \quad , \qquad (2.2.8)$$

and finally

$$d\varepsilon = T\,ds - p\,d(1/\rho) + (\mu_2-\mu_1)\,dc_2 \quad , \qquad (2.2.9)$$

which is the differential form of the relationship $\varepsilon = f(s,\rho,c_2)$ that holds for all equilibrium states of the molecular mixture. Result (2.2.9) is also known as the Gibbs relation for the mixture. Now comes into play the theory of irreversible processes (cf. the text-book by De Groot and Mazur [6]) which assumes : although a flowing mixture is generally not in a equilibrium state, there exist within small mass elements a state of "local" equilibrium for which the internal energy is the same function $f(s,\rho,c_2)$ of the state variables as in complete equilibrium. Hence, the theory of irreversible processs supposes that the Gibbs relation (2.2.9) still holds for a mass element that one follows along its center of mass motion and consequently

$$T\,ds/dt = d\varepsilon/dt + p\,d(1/\rho)/dt - (\mu_2-\mu_1)\,dc_2/dt$$

If that assumption of local equilibrium is taken for granted, the constraint (2.1.3) combined with (2.1.2) and the mass conservation (1.1.1) can be restated as

$$(\Pi + pI) : \nabla\mathbf{v} + T\,\nabla.\mathbf{H} - \nabla.\mathbf{Q} - (\mu_2-\mu_1)\,\rho\,dc_2/dt \geq 0$$

The time-derivative appearing in the above inequality can be eliminated with the mass conservation of component 2 written in (1.1.4) and which can be transformed into

$$\rho\,dc_2/dt = -\,\nabla.\mathbf{J}_2 \qquad (2.2.10)$$

where $\mathbf{J}_2 = \rho_2(\mathbf{v}_2-\mathbf{v})$ is the relative mass flux of species 2 (note that $\mathbf{J}_1 = -\mathbf{J}_2$). Taking that evolution equation into account, the above inequality appears as a sum of terms representing the dissipation rate in the molecular mixture. One must strive to present the result in the form $\Sigma X_m Y_m \geq 0$ where X_m is a flux linked to the mixture motion and Y_m is some quantity vanishing in equilibrium, so that the dissipation rate is automatically zero if the equilibrium conditions are satisfied. That requirement is fulfilled if the entropy and energy fluxes are related as

$$\mathbf{Q} = T\mathbf{H} + (\mu_2-\mu_1)\,\mathbf{J}_2 \qquad (2.2.11)$$

The dissipation rate now appears as

$$(\Pi+pI) : \nabla\mathbf{v} - \mathbf{H}.\nabla T - \mathbf{J}_2.\nabla(\mu_2-\mu_1) \geq 0 \quad . \qquad (2.2.12)$$

At this point, the conclusion is that any model expression proposed for the fluxes $\Pi+pI$, $\mathbf{H}$ and $\mathbf{J}_2$ should satisfy the above inequality. Moreover, those three fluxes vanish in complete equilibrium and are likely to be small for small departures from equilibrium. The simplest answer is to suppose that the fluxes are proportional to the Y_m, i.e. to $\nabla\mathbf{v}$, ∇T and $\nabla(\mu_2-\mu_1)$. It can be shown [6] that for isotropic mixtures, $\Pi+pI$ depends on $\nabla\mathbf{v}$ only, while $\mathbf{H}$ and $\mathbf{J}_2$ can depend on both ∇T and $\nabla(\mu_2-\mu_1)$. We will not insist on the stress tensor (which is newtonian, with a viscosity that may depend on concentration) to focuss on the entropy and mass fluxes. The most general expressions that satisfy the entropy inequality are

$$\mathbf{J}_2 = -A[\nabla(\mu_2-\mu_1) + s^*\nabla T] \qquad (2.2.13)$$

and

$$\mathbf{H} = -B\,\nabla T + s^*\mathbf{J}_2 \quad , \qquad (2.2.14)$$

where A and B are two positive transport coefficients while s^*, with the dimension of an entropy per unit mass, can have <u>any value</u>. Since the chemical potentials are functions of pressure, temperature and concentration, the above expression of $\mathbf{J}_2$ can be rewritten as

$$\rho_2(\mathbf{v}_2-\mathbf{v}) = -\rho D(\nabla c_2 + d_T\,\nabla T/T + d_p\,\nabla p/p) \qquad (2.2.15)$$

with a concentration-diffusion coefficient defined as

$$\rho D = A\,\partial(\mu_2-\mu_1)/\partial c_2 = A\,\partial^2\Delta g/\partial c_2^2 \qquad (2.2.16)$$

while d_T and d_p are the coefficients of thermo-diffusion and baro-diffusion respectively. A comment should be done at that point. The diffusion coefficient is expected to be positive. Since both A and ρ are positive, this means that $\partial^2\Delta g/\partial c_2^2$ must be positive. In fact, this is one of the conditions of stability of the mixture. If this condition fails (for some concentration range) while Δg is still negative, the mixture will generally transform into two homogeneous mixtures of different concentrations, what is called a demixtion phenomenon. In fact, a negative diffusion coefficient means that $\mathbf{v}_2$-$\mathbf{v}$ will be directed along ∇c_2, hence the molecules of species 2 will move towards regions of the flow where species 2 is already in excess, a basically unstable situation which favours demixtion. Result (2.2.15) for molecular mixtures is to be compared with result (1.3.4) for slowly moving multi-phase mixtures. A more complete comparison is presented in the next section.

2.3. *Colloidal suspensions*

When the suspended particules have a size in the range between a few nanometers and a few micrometers, one speaks of a colloidal suspension. Examples of such suspensions of supramolecular particles are microemulsions, polymer solutions, black ink, paint etc... Some of these suspensions are thermodynamically stable ($\Delta g<0$), others are stabilized with the help of some additives (as for instance with surfactants, cf. the lectures by C. Maldarelli in this volume). Similarly to molecular mixtures, the relative motion between the particles and the fluid (or the solute and the solvant) can be the result of inhomogeneities of concentration, temperature and pressure (cf (2.2.15)). In addition, the larger size of the particles, their inertia, can affect that motion. Moreover, as is the case for some polymer solutions, the deformability of the suspended particles will certainly interfere with the relative and global motions. We postpone that problem of deformability to the next section and focuss here to the consequences of the increased particle inertia. A better account of the kinetic energy of the two components is the main improvement to achieve. The simplest expression for the kinetic energy is $c_1v_1^2/2 + c_2v_2^2/2$ for the unit mass of the mixture. That kinetic energy can be rewritten as the energy of the global motion $\mathbf{v}$ plus the energy of the relative motion $\mathbf{v}_n$-$\mathbf{v}$. According to definition (2.1.1), the later is considered as a part of the internal energy and more generally, as a part of any thermodynamic potential such as the free-enthalpy. One is thus led to modify (2.2.1) into

$$g = c_1\mu_1^\circ + c_2\mu_2^\circ + \Delta g + c_1(v_1-v)^2/2 + c_2(v_2-v)^2/2 \quad . \qquad (2.3.1)$$

This means that g is a function of the relative velocity between the two components, besides the usual state variables of a molecular mixture. One cannot consider the relative velocity as a true state variable because it vanishes in equilibrium. However, it is a quantity that must be taken into account to describe non-equilibrium situations. That kind of variable is called an <u>internal variable</u> and a recent review [7] presents a panel of the many fields of continuum physics in which the introduction of internal variables proved to be useful. What are the consequences of the dependence of g on the relative velocity ? While the definitions of the specific entropy and specific volume are similar to those for molecular mixtures (cf. (2.2.4) and (2.2.5)), the definition of the chemical potentials are slightly modified into

$$\mu_n^* = \mu_n + (v_n-v)^2/2 \quad , \qquad (2.3.2)$$

with μ_n defined as in (2.2.3). One can check that $g = c_1\mu_1^* + c_2\mu_2^*$ and that the definition (2.2.6) of the internal energy leads to a Gibbs relation depending on the relative velocity $\mathbf{w} = \mathbf{v}_2$-$\mathbf{v}_1$ as

$$d\varepsilon = T\,ds - p\,d(1/\rho) + (\mu_2^*-\mu_1^*)\,dc_2 + c_1c_2\mathbf{w}.d\mathbf{w} \quad . \qquad (2.3.3)$$

The different steps for obtaining the dissipation rate are similar to those used for molecular mixtures. Taking the conservation laws and (2.2.10) into account, one arrives at

$$(\Pi+pI):\nabla\mathbf{v} - \mathbf{H}.\nabla T - \mathbf{J}_2.[\,d\mathbf{w}/dt + \nabla(\mu_2^*-\mu_1^*)] + \nabla.[T\mathbf{H} - \mathbf{Q} + (\mu_2^*-\mu_1^*)\mathbf{J}_2] \quad \geq 0$$

As will be seen in the next chapter, the mean relative velocity plays an important role in the entropy and momentum fluxes of multi-phase and colloidal mixtures. According to the general expressions (3.3.14) and (3.6.15) one is led to introduce τ and $\mathbf{h}$ defined as

$$\Pi + pI = -\,\mathbf{J}_2\otimes\mathbf{w} + \tau$$

and (2.3.4)

$$\mathbf{H} = (s_2-s_1)\mathbf{J}_2 + \mathbf{h}$$

The entropy inequality is now transformed into

$$\tau:\nabla\mathbf{v} - \mathbf{h}.\nabla T - \mathbf{J}_2.[d_2\mathbf{v}_2/dt - d_1\mathbf{v}_1/dt + \nabla(\mu_2-\mu_1) + (s_2-s_1)\nabla T\,] +$$
$$+\nabla.[T\mathbf{H} - \mathbf{Q} + (\mu_2^*-\mu_1^*)\mathbf{J}_2] \geq 0 \qquad (2.3.5)$$

As noticed in the first chapter, the mechanical dissipation rate depends on the gradient of the volume averaged velocity $\mathbf{u}$ defined in (1.1.11), and <u>not</u> on $\nabla\mathbf{v}$. This can be easily taken into account if the energy flux of the suspension is written as

$$\mathbf{Q} = T\,\mathbf{H} + (\mu_2^*-\mu_1^*)\,\mathbf{J}_2 + (\mathbf{v}-\mathbf{u}).\tau \quad . \qquad (2.3.6)$$

Then, the entropy inequality becomes

$$\tau:\nabla\mathbf{u} - \mathbf{h}.\nabla T - \mathbf{J}_2.[d_2\mathbf{v}_2/dt - d_1\mathbf{v}_1/dt + \nabla(\mu_2-\mu_1) + (s_2-s_1)\nabla T + (\phi_1/\rho_1 - \phi_2/\rho_2)\,\nabla.\tau\,] \geq 0 \ .$$

To present that expression as a true dissipation rate one must eliminate the time-derivatives by proposing model equations for the momentum balance of each component. There are many equivalent ways to write these equations but the above inequality suggests that the most <u>convenient</u> one is

$$\rho_1\,(d_1\mathbf{v}_1/dt - \mathbf{g}) = -\,\phi_1\,\nabla p + \phi_1\,\nabla.\tau - \mathbf{f} - \mathbf{F}_{th} \qquad (2.3.7)$$

$$\rho_2\,(d_2\mathbf{v}_2/dt - \mathbf{g}) = -\,\phi_2\,\nabla p + \phi_2\,\nabla.\tau + \mathbf{f} + \mathbf{F}_{th} \qquad (2.3.8)$$

where the thermodynamical force $\mathbf{F}_{th}$ is defined as

$$\mathbf{F}_{th} = -\rho\, c_1 c_2 [\nabla(\mu_2-\mu_1) + (s_2-s_1)\,\nabla T - (\phi_2/\rho_2 - \phi_1/\rho_1)\,\nabla p] \quad . \qquad (2.3.9)$$

The interpretation of the momentum equations (2.3.7) and (2.3.8) is staightforward : everything happens as if each particle (index 2 represents the particulate phase) was submitted to the average force $-\nabla p + \nabla.\tau$ (note that this averaged force is long-ranged as compared to the particle size) and that the mean value of the remaining forces was represented by $(\mathbf{f} + \mathbf{F}_{th})/\phi_2$. With the above momentum equations, the dissipation rate now appears in the quite simple form

$$\tau : \nabla\mathbf{u} - \mathbf{h}.\nabla T - \mathbf{w}.\mathbf{f} \geq 0 \quad . \qquad (2.3.10)$$

That inequality must be satisfied by any model expressions proposed for τ, $\mathbf{h}$ and $\mathbf{f}$. Results (2.3.6) to (2.3.10) are representative of colloidal suspensions. They also hold for two-phase mixtures with negligible differences between the average pressure and temperature of the two phases (see e.g. [8] for the incidence of pressure and temperature differences). Let us now comment a little bit on the thermodynamic force $\mathbf{F}_{th}$. This force is a generalisation of the osmotic force met in the first chapter. It is proportional to the gradient of the chemical potential difference, corrected by a pressure force which is necessary not to count twice the Archimede's thrust already present in the momentum equations, and by a thermal force which accounts for the role of the relative velocity in the entropy flux (cf. (2.3.4)). According to the thermodynamic results (2.2.8) and (2.2.3) to (2.2.5), one can rewrite (2.3.9) as

$$\mathbf{F}_{th} = -\rho c_1 c_2\, [\partial^2\Delta g/\partial c_2^2\ \nabla c_2 + \partial\mu/\partial T\ \nabla T + \partial\mu/\partial p\ \nabla p] \qquad (2.3.11)$$

with

$$\mu = c_1\partial\Delta g_1/\partial c_2 + c_2\partial\Delta g_2/\partial c_2 \quad .$$

Hence, the thermodynamic force depends on ∇c_2 , ∇T and ∇p with coefficients that depend on the expression of the free-enthalpy of mixing. The ∇p term vanishes for two incompressible phases and it is likely to be very small for all colloidal mixtures, because of the usually weak dependence of Δg on pressure. The ∇T term is interesting because it suggests the possibility of acting on the particle motion with a temperature gradient (we return later to the so-called Soret effect). The ∇c_2 term was already commented upon in the previous section and we now discuss its connextion with the osmotic pressure. When the two components of the mixture are incompressible, Δg is independent of pressure and the mass fraction c_2 is a function of the volume fraction ϕ_2 only. Then, one introduces a free-enthalpy per unit volume Δg^* defined as

$$\Delta g^* = \rho\,\Delta g = \rho_1\,\Delta g_1(\phi_2,T) + \rho_2\,\Delta g_2(\phi_2,T) \quad . \qquad (2.3.12)$$

The incompressibility of both components implies $dc_2/c_1c_2 = d\phi_2/\phi_1\phi_2$ and (2.3.11) becomes

$$\mathbf{F}_{th} = -\ \phi_1\phi_2\,[\partial^2\Delta g^*/\partial\phi_2^2\ \nabla\phi_2 + \partial\mu^*/\partial T\ \nabla T] \qquad (2.3.13)$$

with

$$\mu^* = \rho_1 \partial \Delta g_1 / \partial \phi_2 + \rho_2 \partial \Delta g_2 / \partial \phi_2 .$$

When the temperature of the suspension is a constant, a still simpler expression holds

$$\mathbf{F}_{th} = - \phi_1 \nabla P_{osm} \tag{2.3.14}$$

whith an osmotic pressure $P_{osm}(\phi_2)$ defined as

$$P_{osm} = \phi_2 \, \partial \Delta g^* / \partial \phi_2 - \Delta g^* . \tag{2.3.15}$$

It is left as an exercise to prove that the osmotic pressure is nothing but a disguised way to represent the chemical potential of species 1 since $P_{osm} = - \rho_1°(\mu_1 - \mu_1°)$. As an example, the osmotic pressure of hard-spheres particles is

$$P_{osm} = n_2 \, k_B T \, [1 + 2\phi_2(2-\phi_2) / (1-\phi_2)^3] \quad , \tag{2.3.16}$$

where $n_2 = 3\phi_2 / 4\pi a^3$ is the number of particles per unit volume. One can check that if $\mathbf{F}_{th}$ is given by (2.3.14) and $\mathbf{f} = - \phi_1 R(\mathbf{v}_2 - \mathbf{v}_1)$ (with a positive friction coefficient R so as to satisfy (2.3.10)), the momentum equations (2.3.7) and (2.3.8) merge into the equations (1.3.6) and (1.3.7), except for the contact or collision force. When a temperature gradient occurs, the thermodynamic force will display a contribution directed along ∇T (cf.(2.3.13)), but most interestingly, the friction force **f** can also depend on ∇T . In fact, if the reduced entropy flux **h** happens to involve the relative velocity as

$$\mathbf{h} = - B \nabla T + \phi_1 \phi_2 s^{**} \mathbf{w}$$

the entropy inequality (2.3.10) will force **f** to appear as

$$\mathbf{f} = - \phi_1 R \mathbf{w} - \phi_1 \phi_2 s^{**} \nabla T$$

where B and R are positive transport coefficients while s^{**} (with the dimension of a specific entropy), is an arbitrary scalar. The total ∇T force appearing in the interphase force $\mathbf{f} + \mathbf{F}_{th}$ (the Soret force) will depend on $s^{**} + \partial \mu^* / \partial T$. Generally speaking, the coefficient s^{**} will depend on the type and strength of the flow considered and is quite difficult to compute, at variance with $\partial \mu^* / \partial T$ which depends on the thermostatics properties of the mixture only. Moreover, if the dependence of **h** on **w** is no longer isotropic, the scalar s^{**} is transformed into a tensor. Note finally that if **h** happens to depend on $\nabla \mathbf{u}$, the entropy inequality will force τ to include thermal stresses depending on the temperature gradient. The number of different possibilities offered by (2.3.10) is unexpectedly large.

2.4. *Suspensions of deformable particles.*

Examples of deformable particles are polymer chains, red cells, drops etc... . When submitted to a non-uniform flow, their shape will be modified but the deformation will be limited because of forces that tend to restore the equilibrium shape. These forces come from the surface tension for drops, from the membrane elasticity for red cells and are of entropic origin for polymer chains. The intensity of the restoring force may depend on the magnitude of the deformation, and follows from some elastic energy. In fact, an elastic free-enthalpy Δg_2^{el} will be added to the free-enthalpy of mixing. If index 2 stands for the particulate phase and if the particle shape is represented symbolically by C, the total free-enthalpy will appear as in (2.3.1), but with Δg now depending on the particle deformation as

$$\Delta g = c_1 \Delta g_1(c_2,p,T) + c_2 \Delta g_2(c_2,p,T) + c_2 \Delta g_2^{el}(c_2,p,T,C) \qquad (2.4.1)$$

with

$$\partial\Delta g/\partial C \mid_{C=C_{eq}} = 0 \quad , \qquad (2.4.2)$$

because Δg must display a minimum when the particles have their equilibrium shape C_{eq}. The Gibbs relation will display that dependence on deformation in the form

$$d\varepsilon = T\,ds - p\,d(1/\rho) + (\mu_2^*-\mu_1^*)\,dc_2 + c_1c_2\mathbf{w}.d\mathbf{w} + \partial\Delta g/\partial C\,dC \quad . \qquad (2.4.3)$$

The particle deformation (represented by $C-C_{eq}$) provides us with a second example of internal variable. When associated with equations (2.2.10), (2.3.7) and (2.3.8), the entropy constraint (2.1.3) now appears as

$$\tau : \nabla\mathbf{u} - \mathbf{h}.\nabla T - \mathbf{w}.\mathbf{f} - \rho\,\partial\Delta g/\partial C\,d_2C/dt + \nabla.[T\,\mathbf{H} - \mathbf{Q} + (\mu_2^*-\mu_1^*)\,\mathbf{J}_2 + (\mathbf{v}-\mathbf{u}).\tau] \geq 0 \qquad (2.4.4)$$

provided the thermodynamic force appearing in the momentum equations (2.3.7) and (2.3.8) is now defined as

$$\mathbf{F}_{th} = -\rho c_1c_2\,[\partial^2\Delta g/\partial c_2^2\,\nabla c_2 + \partial\mu/\partial T\,\nabla T + \partial\mu/\partial p\,\nabla p + c_2\partial^2\Delta g_2^{el}/\partial c_2\partial C\,\nabla C] \qquad (2.4.5)$$

with $\qquad \mu = c_1\partial\Delta g_1/\partial c_2 + c_2\partial(\Delta g_2 + \Delta g_2^{el})/\partial c_2$

The salient new feature is the possibility for $\mathbf{F}_{th}$ to depend on the gradient of deformation. To transform the above inequality (2.4.4) into a true dissipation rate, one must answer two questions:

i) what is the best candidate to represent the particle shape ?

ii) what is the evolution equation for the chosen C ?

Generally speaking, the choice of C will depend on the equilibrium shape, and on the kind of deformation the particle is likely to suffer. For tiny bubbles which hardly deform but may inflate or shrink, the relevant C is the bubble radius. In case of strong deformations which transform the particles into straight filaments, the best choice for C is a vector, somewhat similar to the orientation vector used in the first chapter for rigid fibers. In case of particles made from an elastic solid and with a spherical equilibrium shape, the deformation is conveniently represented by a symmetric second-order tensor C_{ij} representing the average deformation gradient inside the particles. In case of long polymer chains, the best C is also a second -order tensor but representing the average $\langle\langle r_i r_j \rangle\rangle$, where **r** is the vector joining the positions of the first and last monomers of the chain. In case of quasi-spherical capsules protected by a membrane, two different second-order tensors are necessary to describe the deformation (see the lectures by D. Barthes-Biesel in this volume). In case of drops, the use of a second-order tensor is restricted to relatively small deformations like those obtained in a weak shear flow. For large shear or elongational flows, the simultaneous use of higher-order tensors is necessary. Roughly speaking, this is because the shape of the deformed drop is often more complicated than the simple ellipsoidal shape described by C_{ij} [9]. The difficulty to find a pertinent C is still more acute in case of particles which, like red-cells in blood, display a non-spherical equilibrium shape. In what follows, we focuss on the case of a second-order tensor C_{ij} as the pertinent deformation variable [10], and consider its equation of evolution.

Since the very definition of C_{ij} is not the same for elastic particles and polymer solutions, one cannot expect a universal evolution equation. However, our only aim here is to emphasize the general incidence of the particle deformability on the global and relative motions of the suspension. For that purpose, we will devise some idealized equation for a non-dimensional $C_{ij}(\mathbf{x},t)$. We assume that the particle is affinely deformed by the gradient of some velocity **V** [11], and is submitted to a restoring mechanism that vanishes in the equilibrium state. Then, since the particles move wih velocity $\mathbf{v}_2$, one writes

$$d_2C_{ij}/dt = C_{ik}\,\partial V_j/\partial x_k + C_{jk}\,\partial V_i/\partial x_k - R_{ij} \quad . \qquad (2.4.6)$$

If the suspension is a very dilute one, the particle deformation depends on the velocity field of the surrounding fluid and **V** is likely to be $\mathbf{v}_1$. If the suspension is concentrated enough for the particles to become entangled, they will form a kind of lattice or gel, the deformation of a particle will be bound to the deformation of the lattice as a whole, and the above **V** is likely to be $\mathbf{v}_2$. To encompass all possible cases, let us write

$$\mathbf{V} = \mathbf{u} + \alpha\,(\mathbf{v}_2 - \mathbf{v}_1) \qquad (2.4.7)$$

where **u** is the volume flux of the suspension, while α is a parameter which, according to the above remarks, mainly depends on the particle concentration. It is now easy to obtain from (2.4.6) and (2.4.7)

$$\rho\,\partial\Delta g/\partial C_{ij}\; d_2C_{ij}/dt = \tau^{el} : \nabla\mathbf{u} - \alpha\mathbf{w}.\nabla.\tau^{el} + \nabla.(\alpha\mathbf{w}.\tau^{el}) - \rho\,R_{ij}\,\partial\Delta g/\partial C_{ij} \quad ,$$

where τ^{el} is an "elastic" stress tensor depending on the particle deformation and defined as

$$(\tau^{el})_{ij} = 2\,\rho\,C_{kj}\,\partial\Delta g/\partial C_{ik} \quad . \qquad (2.4.8)$$

Many expressions of Δg are possible but its dependance on the deformation C_{ij} must guarantee that Δg is a true scalar. The general condition is $C_{kj}\, \partial\Delta g/\partial C_{ik} = C_{ki}\, \partial\Delta g/\partial C_{jk}$ and consequently the elastic stress tensor is <u>symmetric</u>. The energy flux of a suspension of deformable particles will be defined as

$$\mathbf{Q} = T\,\mathbf{H} + (\mu_2^*-\mu_1^*)\,\mathbf{J}_2 + (\mathbf{v}-\mathbf{u}).\tau - \alpha\mathbf{w}.\tau^{el} \quad , \qquad (2.4.9)$$

and the inequality (II.4.4) is then transformed into its final form

$$(\tau - \tau^{el}) : \nabla\mathbf{u} - \mathbf{h}.\nabla T - \mathbf{w}.(\mathbf{f} - \alpha\nabla.\tau^{el}) + \rho\, R_{ij}\, \partial\Delta g/\partial C_{ij} \geq 0 \quad . \qquad (2.4.10)$$

The consequences of (2.4.10) are interesting and we end that section by developping some of them. The simplest expression for R_{ij} is proportional (with a positive coefficient) to $\partial\Delta g/\partial C_{ij}$ and (2.4.2) implies that R_{ij} vanishes at equilibrium, as expected. Concerning the stress τ, it will appear as the sum of the elastic stress τ^{el} and a dissipative stress proportional to the symmetric part of $\nabla\mathbf{u}$. Since τ^{el} depends on the deformation and the deformation obeys the evolution equation (2.4.6), all the ingredients are prepared for the mixture to behave as a non-newtonian fluid. Concerning the intercomponent force **f**, the new feature is that it now depends on the particle deformation whenever $\alpha \neq 0$. For instance, inequality (2.4.10) is satisfied with

$$\mathbf{f} = -\phi_1 R\mathbf{w} + \alpha\nabla.\tau^{el} \qquad (2.4.11)$$

The appearance of the elastic tensor τ^{el} in both τ and **f** implies a coupling between the relative and global motion with interesting consequences [12]. The part of the total interphase force $\mathbf{f} + \mathbf{F}_{th}$ which depends on deformation and deformation gradients is obtained from (2.4.5) and (2.4.11) as

$$\mathbf{f} + \mathbf{F}_{th} = \alpha\nabla.\tau^{el} - \rho c_1 c_2^2\,(\partial^2\Delta g_2^{el}/\partial c_2\partial C_{ij})\nabla C_{ij} + \ldots \quad . \qquad (2.4.12)$$

For a suspension of elastic particles or a dilute polymer solution, the elastic energy is proportional to the particle concentration, then Δg_2^{el} is independent of c_2 and the ∇C_{ij} force desappears. However, this is not the case for semi-dilute polymer solutions where the entanglement of the chains results in a elastic energy increasing non-linearly with c_2. The intercomponent force also depends on the elastic stress $\nabla.\tau^{el}$ provided $\alpha \neq 0$. This leads to the interesting possibility of an <u>anisotropic deformation-induced diffusion</u> of the particles. Again, entangled polymer chains are the best candidates to display that effect because the deformation of the polymer network is linked to the gradient of $\mathbf{v}_2$, hence $\alpha = \phi_1$ according to (2.4.7).

3. ENSEMBLE AVERAGING AND THE MAIN BALANCE EQUATIONS

This chapter presents in their detailed form, the main balance laws describing a two-phase mixture of fluid and particles (the extension to multi-phase mixtures is straightforward and briefly sketched at the end of the chapter). The three basic ingredients are the balance laws for the pure phases, the function of presence of a phase in the mixture and the statistical average over all possible configurations. The definitions of average quantities are chosen so as to present the equations for the mixture in a form similar to the well-known equations for a pure fluid. Not to confuse the issue, we will consider fluid-like particles only and delete surface tension phenomena.

3.1 *A mathematical tool : the function of presence*

To get field equations describing the mixture, one needs a tool to express the random occupation of point **x** by one of the two phases, and its evolution with time t [13]. Let χ_p be a function of position and time defined as

$$\chi_p(x,t) = 1 \quad \text{if point } \mathbf{x} \text{ is inside a particle at time t} \quad ,$$
$$\chi_p(x,t) = 0 \quad \text{if point } \mathbf{x} \text{ is inside the fluid at time t} \quad .$$

χ_p is the function of presence of the particulate phase in the mixture. A similar function χ_f can be defined for the fluid phase. Since the two phases occupy the whole available space

$$\chi_f = 1 - \chi_p \tag{3.1.1}$$

and a single function of presence is enough to describe a two-phase mixture. Two important properties of the particles function of presence are

$$\nabla\chi_p = - \mathbf{n}_p\,\delta_I \tag{3.1.2}$$

and

$$\partial\chi_p/\partial t + \mathbf{v}_I.\nabla\chi_p = 0 \tag{3.1.3}$$

where δ_I is the function of presence of the interfaces between the fluid and the particles, $\mathbf{v}_I$ is the velocity of those interfaces and $\mathbf{n}_p$ their unit normal pointing outwards the particles. These results have a rather intuitive explanation : result (3.1.2) says that at some instant t, the spatial derivative of χ_p vanishes everywhere except on the interfaces. Since a gradient is directed towards increasing values of its argument, $\nabla\chi_p$ points towards the particles, in the direction opposite to $\mathbf{n}_p$. As to δ_I, it is a Dirac function located on the interfaces and resulting from the step-like behaviour of χ_p on those interfaces. Result (3.1.3) means that at some given position **x**, the time derivative of χ_p is always null, except when a piece of interface is crossing that point. As a consequence, if one stands in the immediate neighborhood of the interface and move with its local velocity $\mathbf{v}_I$, the value of χ_p (either one or zero) will never change. The two above results can be combined as

$$\partial \chi_p / \partial t = \mathbf{v}_I \cdot \mathbf{n}_p \delta_I \tag{3.1.4}$$

Note that the normal component only of the interface velocity appears in the time evolution of the particles function of presence. The function of presence depicts a given configuration of the two phases, i.e. some spatial distribution of the fluid and particles over the whole available space. The evolution in time of a configuration is deterministic. However, the initial conditions are never known precisely and one cannot but resort to a statistical average over all configurations resulting from the possible initial ones. Concerning χ_p, such a statistical average will give the probability of presence of the particles at point $\mathbf{x}$ and time t, a quantity that will henceforth be noted as $\phi(x,t)$. Representing the statistical average over configurations by brackets, one will write

$$<\chi_p> = \phi \qquad \text{and} \qquad <\chi_f> = 1 - \phi \tag{3.1.5}$$

while the statistical average of (3.1.2) and (3.1.4) will appear as

$$<\mathbf{n}_p \delta_I> = -\nabla\phi \tag{3.1.6}$$

and

$$<\mathbf{v}_I \cdot \mathbf{n}_p \delta_I> = \partial\phi / \partial t \tag{3.1.7}$$

It is noteworthy that we assumed the commutativity between the average over configurations and the space or time derivatives. That feature of statistical averaging is crucial for all what follows but is not that obvious. We give below some hints regarding its validity together with a more formal derivation of (3.1.2) and (3.1.4).

In Fig.2 is represented the configuration at some instant t of a fluid-particles mixture, together with an arbitrary but fixed volume V which contains a part of the mixture. Some of the particles are fully inside V, some others are fully outside and some are partly inside and partly outside. For any particle which is fully inside, the integral of the unit normal over the (closed) particle surface S_p vanishes. For those particles which happen to be cut by the boundary S of volume V, a closed surface is obtained by associating the part of S_p that is inside V with the intersection of the particle volume V_p with S. As a consequence,

$$\int_{V_p \cap S} \mathbf{n}\, dS + \int_{S_p \in V} \mathbf{n}_p\, dS = 0.$$

Summing the contributions of all the particles which are fully or partially inside V, one gets

$$\int_S \chi_p \mathbf{n}\, dS + \int_V \mathbf{n}_p \delta_I\, dV = 0\,. \tag{3.1.8}$$

It is now easy to transform that result into an integral over the volume V, from which (3.1.2) is deduced. Result (3.1.8) holds for a given configuration depicted by a single χ_p. Performing a statistical average over all possible configurations and noticing that V is fixed in space leads to (3.1.6). To obtain (3.1.4), consider the evolution in time of that part of the volume of a particle which is inside V. Since V is fixed, its boundary S is motionless and

$$d/dt \int_{V_p} dV = \int_{S_p \in V} \mathbf{v_I} . \mathbf{n}_p \, dS$$

Summing the contributions of all particles fully or partially inside V, one gets

$$d/dt \int_V \chi_p \, dV = \int_V \mathbf{v_I} . \mathbf{n}_p \, \delta_I \, dV \qquad (3.1.9)$$

When transforming the above equality into an integral over the fixed volume V one recovers (3.1.4). And when performing the statistical average of (3.1.9), result (3.1.7) is easily deduced.

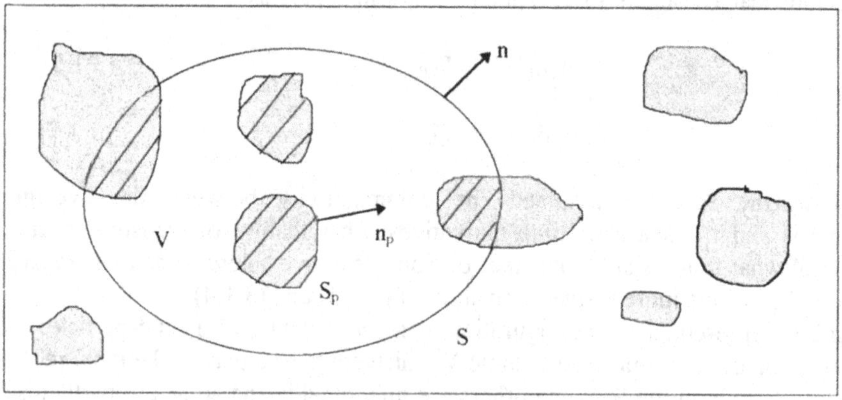

Fig.2 : snapshot of a fluid-particles mixture showing the configuration at some instant t in the laboratory frame. The volume V, which contains a part of the mixture, is motionless in that frame but the mixture configuration is changing with time.

3.2 *The mass balance*

The expression of mass conservation in a region of space occupied by a single phase is well-known. Using an upper index ° for local-instant values of the variables, the mass conservation of pure phase n (n = p for the particles and n = f for the fluid phase) is expressed as

$$\partial \rho_n^{\circ} / \partial t + \nabla . \rho_n^{\circ} \mathbf{v}_n^{\circ} = 0 \quad . \qquad (3.2.1)$$

Equation (3.2.1) is the result of some averaging over tiny volumes containing many molecules of phase n. The averaging has smoothed the molecular structure and presented phase n as a continuum. We want now to derive an expression of the mass conservation when phase n is inside a mixture and when the mixture itself is considered as a continuum.

In a first step, we will obtain an equation taking into account the random occupation of any point of the mixture by one of the two phases. To get it, we just have to multiply the above equation by the function of presence of phase n and to make use of results (3.1.2) and (3.1.4) to obtain

$$\partial \chi_n \rho_n^\circ / \partial t + \nabla . \chi_n \rho_n^\circ \mathbf{v}_n^\circ = \rho_n^\circ (\mathbf{v}_I - \mathbf{v}_n^\circ) . \mathbf{n}_n \delta_I . \qquad (3.2.2)$$

This is an exact result but it is useless because we cannot give any explicit expressions for δ_I and χ_n. Firstly, because the spatial configuration of the interfaces is terribly complicated, and secondly because the initial configuration is not known with precision. We are thus somewhat forced to consider (3.2.2) from a statistical point of view and to take its average over the various possible configurations. Benefitting from the commutativity of that average with the time and spatial derivatives, one deduces

$$\partial \rho_n / \partial t + \nabla . \rho_n \mathbf{v}_n = M_n \qquad (3.2.3)$$

with the definitions

$$\rho_n = < \chi_n \rho_n^\circ > \qquad (3.2.4)$$

$$\rho_n \mathbf{v}_n = < \chi_n \rho_n^\circ \mathbf{v}_n^\circ > \qquad (3.2.5)$$

and

$$M_n = < \rho_n^\circ (\mathbf{v}_I - \mathbf{v}_n^\circ) . \mathbf{n}_n \delta_I > . \qquad (3.2.6)$$

M_n is the mass production rate of phase n, and it can be positive or negative. There is no mass which can accumulate at the interfaces, so that the mass balance at the interfaces between the particles and the fluid is expressed as

$$\rho_p^\circ (\mathbf{v}_I - \mathbf{v}_p^\circ) . \mathbf{n}_p + \rho_f^\circ (\mathbf{v}_I - \mathbf{v}_f^\circ) . \mathbf{n}_f = 0 . \qquad (3.2.7)$$

That boundary condition implies

$$M_p + M_f = 0 . \qquad (3.2.8)$$

The two balance laws (3.2.3) and the link (3.2.8) between the mass exchanges are the continuum equations we were looking for. The definitions (3.2.4) to (3.2.6) are such that the final equation looks very similar to the initial one (3.2.1). In fact, (3.2.3) has a rather intuitive form that could have been anticipated without any averaging procedure (cf.(1.1.4) in the first chapter). However, it is important not to forget the precise definition of "intuitive" quantities. Consider for instance $\rho_p(\mathbf{x},t)$. It is the result of an average over configurations and this means that the contribution to ρ_p of a given configuration is either ρ_p° or zero. If ρ_p° is a constant (incompressible particle) the result of the averaging lies in the range between zero and ρ_p°, and more precisely $\rho_p = \phi \rho_p^\circ$, where $\phi = < \chi_p >$ is the probability of presence of the particles. As to the mean velocity $\mathbf{v}_n$, one must keep in mind its definition (3.2.5), because if the local-instant velocity $\mathbf{v}_n^\circ$ is written as the sum of the mean velocity and a velocity fluctuation

$$\mathbf{v}_n^\circ = \mathbf{v}_n + \mathbf{v}_n' \quad , \tag{3.2.9}$$

then

$$< \chi_n \rho_n^\circ \mathbf{v}_n' > = 0 \tag{3.2.10}$$

while $< \chi_n \mathbf{v}_n' >$ is different from zero except if phase n is incompressible. Let us also remark that the mass exchange rate vanishes if the interface is a non-permeable surface satisfying $\mathbf{v}_I.\mathbf{n}_n = \mathbf{v}_n^\circ.\mathbf{n}_n$ everywhere. Conversely, a non-zero M_n witnesses the presence of chemical reactions or phase transformations between the fluid and the particles. One can benefit from the mutual cancellation of the mass exchanges to obtain from (3.2.3) and (3.2.8) the expression of mass conservation for the whole mixture (cf. (1.1.1))

$$\partial\rho/\partial t + \nabla.\rho\mathbf{v} = 0 \ , \tag{3.2.11}$$

with the following definitions for the mass per unit volume and the mass-weighted velocity of the mixture (cf.(1.1.5) and (1.1.6))

$$\rho = \rho_p + \rho_f \tag{3.2.12}$$

$$\rho\mathbf{v} = \rho_p\mathbf{v}_p + \rho_f\mathbf{v}_f \ . \tag{3.2.13}$$

In the study of mixtures, it has been customary to give $\mathbf{v}$ a prominent role and to measure the velocities relative to it. For that purpose is introduced the relative velocity $\mathbf{w}_n$ defined as

$$\mathbf{w}_n = \mathbf{v}_n - \mathbf{v} \tag{3.2.14}$$

and verifying

$$\rho_p\mathbf{w}_p + \rho_f\mathbf{w}_f = 0 \ . \tag{3.2.15}$$

Another important quantity is the mass fraction c of the particles which is defined as

$$c = \rho_p / \rho \quad . \tag{3.2.16}$$

The fluid mass fraction is $c_f = 1 - c$. With the above definitions, a convenient way to express the two mass balance laws is

$$d\rho/dt = - \rho \nabla.\mathbf{v} \tag{3.2.17}$$

and

$$\rho \ dc/dt = M_p - \nabla.\rho c\mathbf{w}_p \ , \tag{3.2.18}$$

where $d/dt \equiv \partial/\partial t + \mathbf{v}.\nabla$ represents the convective time-derivative following $\mathbf{v}$,

3.3 *The momentum balance*

The same procedure, used above for the mass balance, allows the transformation of the momentum balance of pure phase n written as

$$\partial \rho_n^\circ \mathbf{v}_n^\circ / \partial t + \nabla . \rho_n^\circ \mathbf{v}_n^\circ \otimes \mathbf{v}_n^\circ = \nabla . \sigma_n^\circ + \rho_n^\circ \mathbf{g}_n^\circ \tag{3.3.1}$$

into the average momentum balance for phase n in the mixture

$$\partial \rho_n \mathbf{v}_n / \partial t + \nabla . \rho_n \mathbf{v}_n \otimes \mathbf{v}_n = \nabla . \sigma_n + \mathbf{P}_n + \rho_n \mathbf{g}_n \tag{3.3.2}$$

or equivalently

$$\rho_n \, d_n \mathbf{v}_n / dt = \nabla . \sigma_n + \mathbf{P}_n - M_n \mathbf{v}_n + \rho_n \mathbf{g}_n \tag{3.3.3}$$

where

$$d_n/dt \equiv \partial / \partial t + \mathbf{v}_n . \nabla \tag{3.3.4}$$

is the convective time-derivative following the mean velocity of phase n. The definitions of M_n, ρ_n and $\mathbf{v}_n$ have been given previously. The other quantities appearing in (3.3.2) or (3.3.3) are defined as

$$\sigma_n = \langle \chi_n \sigma_n^\circ \rangle - \langle \chi_n \rho_n^\circ \mathbf{v}_n' \otimes \mathbf{v}_n' \rangle \tag{3.3.5}$$

$$\mathbf{P}_n = \langle \sigma_n^\circ . \mathbf{n}_n \delta_I \rangle + \langle \rho_n^\circ \mathbf{v}_n^\circ (\mathbf{v}_I - \mathbf{v}_n^\circ) . \mathbf{n}_n \delta_I \rangle \tag{3.3.6}$$

and

$$\rho_n \mathbf{g}_n = \langle \chi_n \rho_n^\circ \mathbf{g}_n^\circ \rangle \tag{3.3.7}$$

Except for the interphase force $\mathbf{P}_n$ representing the momentum exchange between the fluid and the particles, the structure of (3.3.2) or (3.3.3) is very similar to that of (3.3.1). Again, this is a consequence of the definitions (3.3.5) to (3.3.7). The price to be paid is that the average stress σ_n must include a contribution from the velocity fluctuations which is quite similar to the Reynolds stress in turbulence.

The momentum balance at the interface between fluid and particles appears as

$$\sum_n \left[\sigma_n^\circ . \mathbf{n}_n + \rho_n^\circ \mathbf{v}_n^\circ (\mathbf{v}_I - \mathbf{v}_n^\circ) . \mathbf{n}_n \right] = 0 \,, \tag{3.3.8}$$

provided surface tension forces can be neglected. If this assumption is taken for granted, the above boundary condition implies

$$\mathbf{P}_p + \mathbf{P}_f = 0 \,. \tag{3.3.9}$$

Due to that mutual cancellation of the momentum exchanges, one can add the two phasic momentum balances and obtain the balance for the whole mixture as

$$\sum_n \left(\rho_n \, d_n \mathbf{v}_n / dt + M_n \mathbf{v}_n \right) = \nabla . \sigma + \rho \mathbf{g} \tag{3.3.10}$$

or

$$\rho \, dv/dt = \nabla . \Pi + \rho \mathbf{g} \tag{3.3.11}$$

where the global external force and stress tensors of the mixture are defined as

$$\rho \mathbf{g} = \rho_p \mathbf{g}_p + \rho_f \mathbf{g}_f \tag{3.3.12}$$

$$\sigma = \sigma_p + \sigma_f \tag{3.3.13}$$

and

$$\Pi = \sigma_p + \sigma_f - (\rho_p \rho_f / \rho) \mathbf{w} \otimes \mathbf{w}$$

$$= \sum_n [\langle \chi_n \sigma_n^{\circ} \rangle - \langle \chi_n \rho_n^{\circ} \mathbf{v}_n' \otimes \mathbf{v}_n' \rangle - \rho_n \mathbf{w}_n \otimes \mathbf{w}_n] \; . \tag{3.3.14}$$

In the expression of the total stress tensor Π appears the relative velocity $\mathbf{w}$ defined as

$$\mathbf{w} = \mathbf{v}_p - \mathbf{v}_f = \mathbf{w}_p - \mathbf{w}_f \quad . \tag{3.3.15}$$

It is noteworthy that the momentum balance (3.3.11) was previously written on intuitive grounds in (1.1.2). The main result here is the detailed definition of the total stress tensor Π, which includes a contribution from the relative velocity, as well as contributions from the velocity fluctuations.

3.4 *The moment of momentum balance*

In continuum mechanics, the moment of momentum balance of a pure fluid requires the symmetry of the stress tensor, i.e.

$$(\sigma_n^{\circ})_{ij} = (\sigma_n^{\circ})_{ji} \quad . \tag{3.4.1}$$

For our mixture of fluid and particles that requirement, when associated with definitions (3.3.5), (3.3.13) and (3.3.14), implies

$$(\sigma_n)_{ij} = (\sigma_n)_{ji} \tag{3.4.2}$$

and

$$\Pi_{ij} = \Pi_{ji} \quad . \tag{3.4.3}$$

That symmetry of all stress tensors seems to be in conflict with the many papers on mixtures which use a non-symmetric stress tensor in case of particles acted upon by external couples. In fact, there is no contradiction but merely *different definitions* of the stress tensors. It can be shown that the statistical average of the external force can be written as a sequence of increasing moments of that force over the volume of a test- particle, the center of which is fixed at the point where the average is performed [14, 15]. More precisely,

$$\rho_p \mathbf{g}_p = \langle \chi_p \rho_p^{\circ} \mathbf{g}_p^{\circ} \rangle =$$

$$n_p <<\int \rho_p^\circ \mathbf{g}_p^\circ \, dV>> - \nabla . [\, n_p <<\int \mathbf{r} \otimes \rho_p^\circ \mathbf{g}_p^\circ \, dV>> + \ldots \,] \tag{3.4.4}$$

where n_p is the average number of particles per unit volume and **r** is the position relative to the particle center. The dots in the squared-brackets represent higher moments of the external force. It is clear now that if we define a new particle stress tensor as

$$\sigma_p^* = \sigma_p - n_p <<\int \mathbf{r} \otimes \rho_p^\circ \mathbf{g}_p^\circ \, dV>> + \ldots \tag{3.4.5}$$

and if we rewrite the momentum balance (3.3.2) for the particulate phase as

$$\partial \rho_p \mathbf{v}_p / \partial t + \nabla . \rho_p \mathbf{v}_p \otimes \mathbf{v}_p = \nabla . \sigma_p^* + \mathbf{P}_p + n_p <<\int \rho_p^\circ \mathbf{g}_p^\circ \, dV>> \tag{3.4.6}$$

then, the presence of a couple acting on the particles will give rise to a non-symmetric σ_p^*. Conversely, if we decide to write the momentum balance of particles as in (3.3.2) or (3.3.3), the stress tensor σ_p is always symmetric and the presence of external couples is somewhat hidden in $\rho_p \mathbf{g}_p$.

3.5 *The energy balance*

The total energy per unit mass of phase n is the sum of an internal and a kinetic energy

$$e_n^\circ = \varepsilon_n^\circ + (v_n^\circ)^2 / 2 \tag{3.5.1}$$

and the energy balance for pure phase n is expressed as

$$\partial \rho_n^\circ e_n^\circ / \partial t + \nabla . \rho_n^\circ e_n^\circ \mathbf{v}_n^\circ = \nabla . (\mathbf{v}_n^\circ . \sigma_n^\circ - \mathbf{q}_n^\circ) + \rho_n^\circ \mathbf{g}_n^\circ . \mathbf{v}_n^\circ \tag{3.5.2}$$

where $\mathbf{q}_n^\circ$ is the local-instant heat flux of phase n. The energy balance of phase n in the mixture is deduced from the averaging of (3.5.2) and appears as

$$\partial \rho_n e_n / \partial t + \nabla . \rho_n e_n \mathbf{v}_n = \nabla . (\mathbf{v}_n . \sigma_n - \mathbf{q}_n) + E_n + \rho_n \mathbf{g}_n . \mathbf{v}_n + \rho_n G_n \tag{3.5.3}$$

or

$$\rho_n \, d_n e_n / dt = \nabla . (\mathbf{v}_n . \sigma_n - \mathbf{q}_n) + E_n - M_n e_n + \rho_n \mathbf{g}_n . \mathbf{v}_n + \rho_n G_n \tag{3.5.4}$$

with the following definitions

$$\rho_n e_n = < \chi_n \rho_n^\circ e_n^\circ > \tag{3.5.5}$$

$$\mathbf{q}_n = < \chi_n \mathbf{q}_n^\circ > - < \chi_n \mathbf{v}_n' . \sigma_n^\circ > + < \chi_n \rho_n^\circ (\varepsilon_n^\circ + (v_n')^2 / 2) \mathbf{v}_n' > \tag{3.5.6}$$

$$E_n = < \mathbf{v}_n^\circ . \sigma_n^\circ . \mathbf{n}_n \delta_I > - < \mathbf{q}_n^\circ . \mathbf{n}_n \delta_I > + < \rho_n^\circ e_n^\circ (\mathbf{v}_I - \mathbf{v}_n^\circ) . \mathbf{n}_n \delta_I > \tag{3.5.7}$$

and

$$\rho_n G_n = \langle \chi_n \rho_n^\circ \mathbf{g}_n^\circ . \mathbf{v}_n' \rangle \qquad (3.5.8)$$

Besides the expected contribution from $\mathbf{q}_n^\circ$, the average energy flux $\mathbf{q}_n$ includes contributions mainly due to velocity fluctuations. The energy exchange rate E_n has contributions from the stress, the heat flux and the mass exchange at the interfaces. Due to (3.2.10), the external energy supply G_n desappears if the external force $\mathbf{g}_n^\circ$ does not fluctuate, as is the case for the gravity field. If the role of surface tension can be neglected, the balance of energy at the interfaces between the fluid and the particles is expressed as

$$\sum_n \left[\mathbf{v}_n^\circ . \sigma_n^\circ . \mathbf{n}_n - \mathbf{q}_n^\circ . \mathbf{n}_n + \rho_n^\circ e_n^\circ (\mathbf{v}_I - \mathbf{v}_n^\circ) . \mathbf{n}_n\right] = 0 \qquad (3.5.9)$$

with the immediate consequence

$$E_p + E_f = 0 \quad . \qquad (3.5.10)$$

That mutual cancellation of the energy exchange at the interfaces allows to write the energy balance for the whole mixture as

$$\sum_n (\rho_n \, d_n e_n/dt + M_n e_n) = \nabla.(\mathbf{v}.\sigma - \mathbf{q}) + \rho \, \mathbf{g}.\mathbf{v} + \rho_p \mathbf{w}_p.(\mathbf{g}_p - \mathbf{g}_f) + \rho G^* \qquad (3.5.11)$$

or

$$\rho \, de/dt = \nabla.(\mathbf{v}.\Pi - \mathbf{Q}) + \rho \, \mathbf{g}.\mathbf{v} + \rho_p \mathbf{w}_p.(\mathbf{g}_p - \mathbf{g}_f) + \rho G^* \qquad (3.5.12)$$

which is to be compared with (1.1.3). In the above balance for total energy, σ and Π are the stresses defined in (3.3.13) and (3.3.14), while the total energy ρe of the mixture, the fluctuational energy supply ρG^* and the total energy fluxes $\mathbf{q}$ and $\mathbf{Q}$ are defined as

$$\rho \, e = \rho_p e_p + \rho_f e_f \qquad (3.5.13)$$

$$\rho \, G^* = \rho_p G_p + \rho_f G_f \qquad (3.5.14)$$

$$\mathbf{q} = \Sigma \, (\mathbf{q}_n - \mathbf{w}_n . \sigma_n) \qquad (3.5.15)$$

and

$$\mathbf{Q} = \sum_n \left[\mathbf{q}_n - \mathbf{w}_n . \sigma_n + \rho_n \mathbf{w}_n (e_n - \mathbf{v}.\mathbf{w}_n)\right] \quad . \qquad (3.5.16)$$

It is interesting to present the total energy of phase n as the sum of a kinetic and an internal energy. From (3.5.1) and (3.5.5) one gets

$$e_n = \varepsilon_n + k_n + (\mathbf{v}_n)^2/2 \qquad (3.5.17)$$

with definitions

$$\rho_n \varepsilon_n = \langle \chi_n \rho_n^\circ \varepsilon_n^\circ \rangle \qquad (3.5.18)$$

and

$$\rho_n k_n = \langle \chi_n \rho_n^\circ (\mathbf{v}_n')^2/2 \rangle \quad . \qquad (3.5.19)$$

Note that k_n is the kinetic energy associated with the velocity fluctuations, a kind of "turbulent" kinetic energy. The sum $\varepsilon_n + k_n$ is to be understood as the *effective* internal energy per unit mass of phase n and its evolution equation is deduced from (3.5.4) and (3.3.3) as

$$\rho_n d_n(\varepsilon_n + k_n)/dt = \boldsymbol{\sigma}_n : \nabla \mathbf{v}_n - \nabla.\mathbf{q}_n + E_n^* - M_n(\varepsilon_n + k_n) + \rho_n G_n \quad (3.5.20)$$

with an exchange term deduced from definitions (3.3.6) and (3.5.7)

$$E_n^* = \langle \mathbf{v}_n'.\boldsymbol{\sigma}_n^\circ.\mathbf{n}_n\delta_I \rangle - \langle \mathbf{q}_n^\circ.\mathbf{n}_n\delta_I \rangle + \langle \rho_n^\circ(\varepsilon_n^\circ + v_n'^2/2)(\mathbf{v}_I - \mathbf{v}_n^\circ).\mathbf{n}_n\delta_I \rangle . \quad (3.5.21)$$

and satisfying the sum rule

$$E_p^* + E_f^* = (\mathbf{v}_f - \mathbf{v}_p) . [\mathbf{P}_p - (M_p/2)(\mathbf{v}_f + \mathbf{v}_p)] \quad . \quad (3.5.22)$$

Concerning the energy of the whole mixture , it can be written in the usual form

$$\rho\, e = \rho\, \varepsilon + \rho\, v^2/2 \quad (3.5.23)$$

provided one defines the internal energy ε of the mixture as

$$\rho\, \varepsilon = \rho_p(\varepsilon_p + k_p) + \rho_f(\varepsilon_f + k_f) + \rho c(1-c)w^2/2 \quad . \quad (3.5.24)$$

In other words, for the total energy of the mixture to have an expression similar to that for a pure fluid, we have to include in the definition of the internal energy, not only the kinetic energy of the relative motion (cf. (2.3.1)) but also the fluctuational kinetic energy. With expression (3.5.17) for e_n, it is possible to rewrite the total energy flux $\mathbf{Q}$ as

$$\mathbf{Q} = \sum_n [\mathbf{q}_n - \mathbf{w}_n.\boldsymbol{\sigma}_n + \rho_n \mathbf{w}_n(\varepsilon_n + k_n + w_n^2/2)] . \quad (3.5.25)$$

The relative motion $\mathbf{w}_n$ plays in the total heat flux $\mathbf{Q}$ a role that was already taken into account in the thermodynamic approach of mixtures (cf. chapter two). The evolution of the internal energy of the mixture can be deduced from the evolution (3.5.12) of the total energy , the momentum balance (3.3.11) and the definition (3.5.23). The result is

$$\rho\, d\varepsilon/dt = \Pi : \nabla \mathbf{v} - \nabla.\mathbf{Q} + \rho_p \mathbf{w}_p.(\mathbf{g}_p - \mathbf{g}_f) + \rho G^* \quad . \quad (3.5.26)$$

If the only external force is the gravity field, the last two terms vanish and the above equation is quite similar to that for a pure fluid (cf. also 2.1.2). The main difference is to be found in the expressions (3.3.14) and (3.5.25) for the total stress and total energy flux of the mixture.

3.6 *The entropy balance*

The entropy balance for pure phase n can be written as

$$\partial\rho_n^\circ s_n^\circ/\partial t + \nabla.(\rho_n^\circ s_n^\circ \mathbf{v}_n^\circ + \mathbf{q}_n^\circ/T_n^\circ) = \Delta_n^\circ \qquad (3.6.1)$$

with a positive entropy production rate Δ_n° that appears as

$$T_n^\circ\Delta_n^\circ = (\sigma_n^\circ + p_n^\circ I) : \nabla v_n^\circ - (\mathbf{q}_n^\circ/T_n^\circ) .\nabla T_n^\circ \geq 0 \qquad (3.6.2)$$

where p_n° and T_n° are the pressure and temperature of phase n, while I is the unit tensor. The averaging procedure applied to (3.6.1) leads to the entropy balance for phase n in the mixture

$$\partial\rho_n s_n/\partial t + \nabla.(\rho_n s_n \mathbf{v}_n + \mathbf{h}_n) = \Delta_n + S_n \qquad (3.6.3)$$

or

$$\rho_n\, d_n s_n/dt = -\nabla.\mathbf{h}_n + \Delta_n + S_n - M_n s_n \qquad (3.6.4)$$

with the following definitions

$$\rho_n s_n = < \chi_n \rho_n^\circ s_n^\circ > \qquad (3.6.5)$$

$$\Delta_n = < \chi_n \Delta_n^\circ > \qquad (3.6.6)$$

$$\mathbf{h}_n = < \chi_n \mathbf{q}_n^\circ/T_n^\circ > + < \chi_n \rho_n^\circ s_n^\circ \mathbf{v}_n' > \qquad (3.6.7)$$

and

$$S_n = < \rho_n^\circ s_n^\circ(\mathbf{v}_I - \mathbf{v}_n^\circ).\mathbf{n}_n \delta_I > - < (\mathbf{q}_n^\circ/T_n^\circ).\mathbf{n}_n \delta_I > . \qquad (3.6.8)$$

S_n represents the entropy exchanges through the interfaces. At variance with results (3.2.8), (3.3.9) and (3.5.10), the sum $S_p + S_f$ is not likely to vanish, because of some possible entropy production located on the interfaces. If surface tension can be neglected, the entropy balance at the interfaces will be expressed as

$$\sum_n [\, \rho_n^\circ s_n^\circ(\mathbf{v}_I - \mathbf{v}_n^\circ).\mathbf{n}_n - (\mathbf{q}_n^\circ/T_n^\circ).\mathbf{n}_n \,] = \Delta_I^\circ \qquad (3.6.9)$$

where Δ_I° represents the rate of entropy production per unit area. As a consequence

$$S_p + S_f = \Delta_I \;, \qquad (3.6.10)$$

where

$$\Delta_I = < \Delta_I^\circ \delta_I > . \qquad (3.6.11)$$

The entropy balance for the whole mixture is deduced from (3.6.3) and (3.6.10), and can be presented in a form quite similar to that for a pure fluid (cf. also (2.1.3)), i.e.

$$\rho \, ds/dt = - \nabla.\mathbf{H} + \Delta \qquad (3.6.12)$$

with definitions

$$\rho s = \rho_p s_p + \rho_f s_f \qquad (3.6.13)$$

$$\Delta = \Delta_p + \Delta_f + \Delta_I \qquad (3.6.14)$$

and

$$\mathbf{H} = \mathbf{h}_p + \mathbf{h}_f + (s_p - s_f)\rho_p \mathbf{w}_p = \mathbf{h} + (s_p - s_f)\, \mathbf{J}_p \qquad (3.6.15)$$

The positive value of the total entropy production rate Δ follows from the positivity of Δ_n° and Δ_I°. In the absence of surface tension phenomena, several different (but equivalent) expressions of Δ_I° can be obtained from the boundary conditions (3.3.8), (5.5.9), (3.6.9) and the definition (3.5.1). For instance, with *arbitrary* temperature and velocity T_s and $\mathbf{v}_s$, the interfacial entropy production can be written as

$$T_s \Delta_I^\circ = \sum_n [(T_s s_n^\circ - \varepsilon_n^\circ - (v_n^\circ - v_s)^2/2)\, \rho_n^\circ (\mathbf{v}_I - \mathbf{v}_n^\circ).\mathbf{n}_n + (1 - T_s/T_n^\circ)\mathbf{q}_n^\circ.\mathbf{n}_n + (\mathbf{v}_s - \mathbf{v}_n^\circ).\sigma_n^\circ.\mathbf{n}_n]$$

A particularly convenient choice is $T_s = T_p^\circ$ and $\mathbf{v}_s = \mathbf{v}_p^\circ$. Then, writing $\sigma_n^\circ = - p_n^\circ I + \tau_n^\circ$ and taking into account the definition of the chemical potential $\mu_n^\circ = \varepsilon_n^\circ + p_n^\circ/\rho_n^\circ - T_n^\circ s_n^\circ$, one finally gets

$$T_p^\circ \Delta_I^\circ = [\mu_f^\circ - \mu_p^\circ + s_f^\circ(T_f^\circ - T_p^\circ) + (1/\rho_p^\circ)(p_p^\circ - p_f^\circ) + (v_f^\circ - v_p^\circ)^2/2]\, \rho_p^\circ(\mathbf{v}_I - \mathbf{v}_p^\circ).\mathbf{n}_p$$
$$+ (1 - T_p^\circ/T_f^\circ)\mathbf{q}_f^\circ.\mathbf{n}_f + (\mathbf{v}_p^\circ - \mathbf{v}_f^\circ).\tau_f^\circ.\mathbf{n}_f \quad , \qquad (3.6.16)$$

which shows that the interfacial dissipation rate is bound to the existence of i) a temperature jump across the interface, ii) a jump of tangential velocities and iii) a jump of normal velocities (phase transitions). If those jumps are negligible, the interfacial dissipation rate can be neglected.

3.7 *The extension to multi-phase mixtures*

In a N-phase mixture, one must distinguish the N-1 different types of interfaces that separate one phase from all the others [16]. The function of presence of the interfaces, which was noted δ_I for a two-phase mixture, will now be written as

$$\delta_I = \sum_{m \neq n} \sum \delta_{mn} \quad ,$$

where δ_{mn} concerns the interface between phase n and phase m. The balance equations for phase n inside the mixture will appear exactly as in (3.2.3), (3.3.2), (3.5.3) and (3.6.3). The only difference will concern the exchange terms which will be the sum of N-1 different contributions. For example,

$$M_n = \sum_{m \neq n} M_{mn}$$

with

$$M_{mn} = <\rho_n^{\circ}(\mathbf{v}_I-\mathbf{v}_n^{\circ}).\mathbf{n}_n\, \delta_{mn}>$$

and similarly

$$\mathbf{P}_n = \sum_{m\neq n} \mathbf{P}_{mn} \quad , \quad E_n = \sum_{m\neq n} E_{mn} \quad , \quad S_n = \sum_{m\neq n} S_{mn}$$

with obvious expressions for $\mathbf{P}_{mn}$, E_{mn} and S_{mn}. The first difficulty is the possibility of distinct boundary conditions for each type of interface. However, if the interface between phases m and n has a negligible surface tension and is characterized by an entropy production rate Δ_{mn}, one will have

$$\begin{aligned} M_{mn} + M_{nm} &= 0 \\ \mathbf{P}_{mn} + \mathbf{P}_{nm} &= 0 \\ E_{mn} + E_{nm} &= 0 \\ S_{mn} + S_{nm} &= \Delta_{mn} . \end{aligned}$$

If that property is shared by all types of interfaces present in the mixture, the balance of mass, momentum, energy and entropy for the whole mixture will look exactly as in (3.2.11), (3.3.11), (3.5.12), and (3.6.12) with

$$\rho = \sum_n \rho_n$$

$$\rho\, \mathbf{v} = \sum_n \rho_n\, \mathbf{v}_n$$

$$\rho\, \mathbf{g} = \sum_n \rho_n\, \mathbf{g}_n$$

$$\Pi = \sum_n (\sigma_n - \rho_n \mathbf{w}_n \otimes \mathbf{w}_n)$$

$$\rho\, e = \sum_n \rho_n(\varepsilon_n + k_n + v_n^2/2)$$

$$\rho\, \varepsilon = \sum_n \rho_n(\varepsilon_n + k_n + w_n^2/2)$$

$$\mathbf{Q} = \sum_n [\mathbf{q}_n - \mathbf{w}_n.\sigma_n + \rho_n(\varepsilon_n + k_n + w_n^2/2)\, \mathbf{w}_n]$$

$$\rho\, s = \sum_n \rho_n\, s_n$$

$$\Delta = \sum_n \Delta_n + \sum_{m\neq n}\sum \Delta_{mn}$$

$$\mathbf{H} = \sum_n (\mathbf{h}_n + \rho_n s_n \mathbf{w}_n) .$$

The use of these multiphasic expressions mostly concern three-phase mixtures, like those met in petroleum engineering (oil + water +gas) or in atmospheric science (air + water + vapour).

4. THE FLUCTUATIONAL KINETIC ENERGY AND THE INTERNAL ENERGY

In the third chapter we have seen how to express the balances of mass, momentum, energy and entropy for each of the two phases. Despite of its apparent completeness, this set of equations is not enough for many a suspension. This is the case if the particles have a chaotic translational motion, or if their rotation rate is different from that of the ambient fluid for instance. These motions manifest themselves as fluctuations relative to the average velocity $\mathbf{v}_p$, and they will take part in the fluctuational kinetic energy k_p. These fluctuations of the particles velocity will induce fluctuations in the surrounding phase due to the boundary conditions for the velocity at the interfaces, and one must consider k_f simultaneously. Moreover, when the particles are not rigid ones, some elastic or compressional energy must be taken into account in the internal energy ε_p of the particulate phase. It is thus obvious that many kinds of suspensions will require more than the mere balance of total energy for each phase. In this chapter, we will obtain general results concerning the balance of internal energy and the balance of fluctuational kinetic energy. These general results will be given explicit forms with the study of some specific examples. To avoid cumbersome results, *phase transitions will be discarded* in most of this chapter, and results taking these transitions into account will be briefly presented in the last section.

4.1. *Evolution of the phasic fluctuational kinetic energy*

The fluctuational kinetic energy of phase n was defined in (3.5.19). To obtain its evolution in time, a first step is to introduce the partition (3.2.9) into the momentum balance (3.3.1) to get

$$\rho_n^\circ(\partial v_{ni}/\partial t + v_{nk}\,\partial v_{ni}/\partial x_k) + \rho_n^\circ(\,\partial v'_{ni}/\partial t + v^\circ_{nk}\,\partial v'_{ni}/\partial x_k + v'_{nk}\,\partial v_{ni}/\partial x_k) = \partial\sigma^\circ_{nik}/\partial x_k + \rho_n^\circ g^\circ_{ni}$$

Then, one makes the dot product with $\chi_n v'_{ni}$, performs the statistical average over configurations and, keeping in mind that $<\chi_n\rho^\circ\mathbf{v}'>$ vanishes (cf. (3.2.10)), one finally obtains [16]

$$\rho_n\, d_nk_n/dt \;=\; -<\chi_n\rho^\circ\mathbf{v}'\otimes\mathbf{v}'>:\nabla\mathbf{v}_n - \nabla.<\chi_n\rho^\circ(v'^2/2)\mathbf{v}'> + <\chi_n\mathbf{v}'.(\nabla.\sigma^\circ+\rho^\circ\mathbf{g}^\circ)> \;. \quad (4.1.1)$$

This equation is somewhat similar to the equation of evolution of the fluctuational kinetic energy in a single-phase turbulent fluid. The first term on the right-hand side describes the influence of the gradient of $\mathbf{v}_n$ on the development of k_n, the second one is a term which is expected to diffuse k_n and the third one represents the power developed by the stresses and the external forces. It will prove useful to rewrite (4.1.1) in a slightly different form, by expressing the power generated by the stresses as

$$<\chi_n\mathbf{v}'.(\nabla.\sigma^\circ)> \;=\; \nabla.<\chi_n\mathbf{v}'.\sigma^\circ> - <\chi_n\sigma^\circ:\nabla\mathbf{v}^\circ> + <\chi_n\sigma^\circ>:\nabla\mathbf{v}_n + <\mathbf{v}'_n.\sigma_n^\circ.\mathbf{n}_n\,\delta_I> \;.$$

With that result, (4.1.1) can be transformed into

$$\rho_n\, d_nk_n/dt \;=\; \sigma_n:\nabla\mathbf{v}_n - <\chi_n\sigma^\circ:\nabla\mathbf{v}^\circ> - \nabla.\mathbf{q}_n^{\,k} + E_n^{\,k} + \rho_nG_n \quad (4.1.2)$$

The average stress σ_n and the external power source $\rho_n G_n$ were previously met in (3.3.5) and (3.5.8) respectively. The flux $\mathbf{q}_n^k$ and exchange term E_n^k (the superscript k is to remind their fluctuational origin) are defined as

$$\mathbf{q}_n^k = <\chi_n \rho^\circ (v'^2/2)\mathbf{v}'> - <\chi_n \mathbf{v}'.\sigma^\circ> \tag{4.1.3}$$

and

$$E_n^k = <\mathbf{v}'_n.\sigma_n^\circ.\mathbf{n}_n \delta_I> \quad . \tag{4.1.4}$$

Note that $\mathbf{q}_n^k$ and E_n^k are parts of the energy flux $\mathbf{q}_n$ (3.5.6) and energy exchange E_n^* (3.5.21), and that we managed so as to present (4.1.2) in a form similar to the evolution equation (3.5.20) for the effective internal energy of phase n. The reason will appear soon.

4.2 *Evolution of the phasic internal energy*

From the definition (3.5.1) and the balance equations (3.5.2) and (3.3.1), one can obtain the evolution equation for ε_n°. After averaging, and taking definition (3.5.18) into account, one gets the evolution of the average internal energy of phase n as

$$\rho_n \, d_n\varepsilon_n/dt = <\chi_n \sigma^\circ : \nabla \mathbf{v}^\circ> - \nabla.\mathbf{q}_n^T + E_n^T \tag{4.2.1}$$

where $\mathbf{q}_n^T$ and E_n^T (the superscript T is to remind their thermal origin) are defined as

$$\mathbf{q}_n^T = <\chi_n \mathbf{q}^\circ> + <\chi_n \rho^\circ \varepsilon^\circ \mathbf{v}'> \tag{4.2.2}$$

and

$$E_n^T = - <\mathbf{q}_n^\circ.\mathbf{n}_n \delta_I> \quad . \tag{4.2.3}$$

It is noteworthy that the power $<\chi_n \sigma^\circ : \nabla \mathbf{v}^\circ>$ generated by the "microscopic" stresses acts in opposite ways for ε_n and k_n. In fact, as already mentionned in the third chapter, the sum $\varepsilon_n + k_n$ is advantageously considered as the *effective* internal energy of phase n, and, according to (4.1.2) and (4.2.1), it evolves according to

$$\rho_n \, d_n(\varepsilon_n + k_n)/dt = \sigma_n : \nabla \mathbf{v}_n - \nabla.(\mathbf{q}_n^k + \mathbf{q}_n^T) + E_n^k + E_n^T + \rho_n G_n \, . \tag{4.2.4}$$

When comparing with the former result (3.5.20), one deduces the necessary relations of compatibility (remember the absence of phase transitions)

$$\mathbf{q}_n^k + \mathbf{q}_n^T = \mathbf{q}_n \tag{4.2.5}$$

and

$$E_n^k + E_n^T = E_n^* \, . \tag{4.2.6}$$

It can be checked that these relations are satisfied indeed, when comparing the former definition (3.5.21) with the above definitions of $\mathbf{q}_n^k$, $\mathbf{q}_n^T$, E_n^k and E_n^T. In other words, one can consider the evolution equations for ε_n and k_n as the two basic expressions of the energy balance of phase n, the former result (3.5.20) being a mere consequence of them.

4.3 *Evolution of the total fluctuational energy*

We noticed in the third chapter that, due to the mutual cancellation of exchange terms, the evolution of the quantities describing the whole mixture (ρ, ρv, ρe etc...) was particularly simple. Concerning the fluctuational kinetic energy, the exchange term is represented by E_n^k (cf (4.1.2)). We do not expect the sum $E_p^k + E_f^k$ to vanish but we will see that the evolution of the total fluctuational energy is related in a remarkable way to the stresses and forces that appear in the momentum balance equations. In the absence of phase transitions, the velocity difference $\mathbf{v}_p^\circ - \mathbf{v}_f^\circ$ vanishes at any point of the interfaces, the boundary condition (3.3.8) becomes $\sigma_p^\circ.\mathbf{n}_p + \sigma_f^\circ.\mathbf{n}_f = 0$,and the sum of exchange terms can be written as

$$E_p^k + E_f^k = (\mathbf{v}_p - \mathbf{v}_f) . < \sigma_f^\circ.\mathbf{n}_f \, \delta_I > \quad . \qquad (4.3.1)$$

Adding the evolution equations (4.1.2) for the two phases one gets

$$\rho_p \, d_p k_p/dt + \rho_f \, d_f k_f/dt = \sigma_p : \nabla \mathbf{v}_p + \sigma_f : \nabla \mathbf{v}_f - < \sigma^\circ : \nabla \mathbf{v}^\circ > \\ - \nabla.(\mathbf{q}_p^k + \mathbf{q}_f^k) + (\mathbf{v}_p - \mathbf{v}_f) . < \sigma_f^\circ.\mathbf{n}_f \, \delta_I > + \rho G^* \quad . \qquad (4.3.2)$$

We introduce now the volume-averaged velocity **u** defined in (1.1.11); **u** is often refered to as the volume flux of the mixture. As already noticed in the second chapter, it is ∇u (and not ∇v) which plays the main role in the expression of the dissipation rate of the mixture. For that reason, we now manage so as to express (4.3.2) with that velocity and find

$$\rho_p \, d_p k_p/dt + \rho_f \, d_f k_f/dt = \sigma : \nabla \mathbf{u} - \mathbf{w}.\mathbf{f} - < \sigma^\circ : \nabla \mathbf{v}^\circ > - \nabla.\mathbf{q}^k + \rho G^* \quad . \qquad (4.3.3)$$

where σ is the suspension stress defined in (3.3.13), while the force **f** and energy flux $\mathbf{q}^k$ are defined as

$$\mathbf{f} = < \sigma_f^\circ.\mathbf{n}_p \, \delta_I > + \nabla.\sigma_p - \phi \, \nabla.\sigma \qquad (4.3.4)$$

and

$$\mathbf{q}^k = \sum_n (\mathbf{q}_n^k - (\mathbf{v}_n - \mathbf{u}).\sigma_n) \quad . \qquad (4.3.5)$$

It can be checked that, with σ and **f** instead of σ_n and $\mathbf{P}_n$, the momentum balance (3.3.3) can be rewritten as

$$\rho_p \, (d_p \mathbf{v}_p/dt - \mathbf{g}_p) = \phi \, \nabla.\sigma + \mathbf{f} \qquad (4.3.6)$$

$$\rho_f \, (d_f \mathbf{v}_f/dt - \mathbf{g}_f) = (1 - \phi) \, \nabla.\sigma - \mathbf{f} \quad . \qquad (4.3.7)$$

This new presentation of the momentum balances is particularly convenient because of its close connexion with the evolution (4.3.3) of the total fluctuational energy. All our subsequent discussion concerning the influence of velocity fluctuations on the motion of the two phases will be based on (4.3.3), (4.3.6) and (4.3.7). Another way to look upon (4.3.3) is to rewrite it as

$$\rho_p\, d_p k_p/dt + \rho_f\, d_f k_f/dt + \langle \sigma^\circ : \nabla \mathbf{v}^\circ \rangle - \rho G^* = \sigma : \nabla \mathbf{u} - \mathbf{w}.\mathbf{f} - \nabla.\mathbf{q}^k \quad . \qquad (4.3.8)$$

Suppose that we know (or imagine) explicit expressions for ρG^*, for $\langle \sigma^\circ : \nabla \mathbf{v}^\circ \rangle$ and for the rate of change of k_p and k_f. This means we have some explicit expression for the left-hand side of (4.3.8). Then, suppose that we are able to rearrange the resulting terms so as to present them in a form similar to the right-hand side, i.e. a first term involving ∇u, a second one involving the relative velocity **w** and all the remaining terms expressed as the divergence of some vector. Then, a term to term identification will lead to explicit expressions for σ and **f**. Provided these expressions are compatible with the definitions (3.3.5), (3.3.13) and (4.3.4), this means we have devised an original way to check explicit expressions for the mean momentum equations. That check will be tested in sections 6 to 8.

4.4 *The rate of change of mechanical energy*

This section will discuss the role and significance of $\langle \sigma^\circ : \nabla \mathbf{v}^\circ \rangle$. To begin with, we deduce from (4.2.1) the evolution equation for the total internal energy

$$\rho_p\, d_p \varepsilon_p/dt + \rho_f\, d_f \varepsilon_f/dt = \langle \sigma^\circ : \nabla \mathbf{v}^\circ \rangle - \nabla.(\mathbf{q}_p^T + \mathbf{q}_f^T). \qquad (4.4.1)$$

The vanishing of the sum $E_p^T + E_f^T$ expresses the mutual cancellation of the heat flux across the interfaces and is a consequence of the boundary condition (3.5.9) and (3.3.8) when phase transitions are absent. Equation (4.4.1) expresses the first law of thermodynamics for the whole mixture. It is clear that thermal effects are represented by the flux $\mathbf{q}_p^T + \mathbf{q}_f^T$ while the rate of change of mechanical energy is represented by $\langle \sigma^\circ : \nabla \mathbf{v}^\circ \rangle$. For a liquid phase, σ° is the sum of a pressure and a viscous stress, while for a solid phase, σ° is an elastic stress mainly. This means that $\langle \sigma^\circ : \nabla \mathbf{v}^\circ \rangle$ is itself the sum of i) the rate of change of elastic and compressional energies and ii) the dissipation rate of mechanical energy, noted Φ in all what follows. As a consequence, one can write

$$\langle \sigma^\circ : \nabla \mathbf{v}^\circ \rangle = \langle \chi_p \sigma_p^{el} : \nabla \mathbf{v}_p^\circ \rangle - \langle \chi_f\, p_f^\circ \nabla.\mathbf{v}_f^\circ \rangle + \Phi \ . \qquad (4.4.2)$$

Obviously, the elastic stress σ_p^{el} would be replaced by a pure pressure stress, were the particles made from a compressible fluid material. The evolution equation (4.3.3) can now be rewritten as

$$\rho_p\, d_p k_p/dt + \rho_f\, d_f k_f/dt + \langle \chi_p \sigma_p^{el} : \nabla \mathbf{v}_p^\circ \rangle - \langle \chi_f\, p_f^\circ \nabla.\mathbf{v}_f^\circ \rangle + \nabla.\mathbf{q}^k = \sigma : \nabla \mathbf{u} - \mathbf{w}.\mathbf{f} + \rho G^* - \Phi. \qquad (4.4.3)$$

The first two terms on the right-hand side represent the energy supplied by the internal forces, while ρG^* is the energy supplied by the external forces. It is then possible to give the following interpretation of (4.4.3) : the difference between the total energy supply and the dissipation rate is converted into fluctuational, elastic and compressional energies, and diffused by $\mathbf{q}^k$.

4.5 *Suspensions of rigid particles in a non-compressible fluid*

We now consider special types of suspensions. If the particles are rigid and the surrounding fluid is incompressible, the compressional and elastic energies are eliminated and the general result (4.4.2) is reduced to

$$<\sigma^\circ : \nabla \mathbf{v}^\circ> \;=\; \Phi \qquad (4.5.1)$$

The dissipation rate is due to the viscous forces that appear in the shear motion of the suspension as a whole and in the relative motion between the particles and the fluid. Viscous forces will also damp the fluctuational kinetic energies and the general structure of Φ is

$$\Phi = \tau^\eta : \nabla \mathbf{u} - \mathbf{f}^\eta . w + \Phi^k \qquad (4.5.2)$$

where Φ^k is the part of the dissipation rate that specifically acts on the fluctuational energy (cf. (4.5.3) below) while the superscript η is to remind the viscous origin of the other terms. As to the fluctuational energy, it is also expected to be created by shear and relative motions. But it can also be diffused by the flux $\mathbf{q}^k$ and is dissipated with rate Φ^k . Accordingly, one expects a general evolution equation of the form

$$\rho_p\, d_p k_p/dt + \rho_f\, d_f k_f/dt = \sigma^k : \nabla \mathbf{u} - \mathbf{w}.\mathbf{f}^k - \nabla.\mathbf{q}^k - \Phi^k + \rho G^* . \qquad (4.5.3)$$

In fact, the general features of that equation are suggested by the evolution equation (4.1.1). If we take (4.5.2) and (4.5.3) for granted and introduce them into the left-hand side of (4.3.8), one immediately deduces from a term to term identification

$$\sigma = \tau^\eta + \sigma^k - p\,I \qquad (4.5.4)$$

and

$$\mathbf{f} = \mathbf{f}^\eta + \mathbf{f}^k + \mathbf{f}_\perp \quad , \qquad (4.5.5)$$

where p is an undetermined pressure related to the incompressibility condition $\nabla.\mathbf{u} = 0$ while $\mathbf{f}_\perp$ is an undetermined force orthogonal to the relative velocity. When the above expressions for σ and $\mathbf{f}$ are introduced in the momentum equations (4.3.6) and (4.3.7), one clearly sees the respective role of dissipative effects and velocity fluctuations . The main issue is to justify (4.5.2) and (4.5.3). We will not provide any proof but will examine in the two following sections some special cases which will help to support the general form of these equations.

4.6 *Added-mass forces and stresses*

It will be shown in the present section that *added-mass effects are nothing but the consequence of the dependence of k_f on the mean relative velocity*. Let us consider rigid and spherical particles moving in a non-viscous and incompressible fluid. The relative motion between the particles and the fluid is responsible for velocity fluctuations in the fluid phase (due to the boundary conditions obeyed by the velocity) and we will suppose that all particles move with the same translational velocity $\mathbf{v}_p$. Since the fluid velocity fluctuations not only depend on $\mathbf{w} = \mathbf{v}_p - \mathbf{v}_f$ but also on the volume fraction of the particles, we express the above assumptions with

$$\Phi^k = 0 \quad , \quad k_p = 0$$

and

$$k_f = \phi\,(1-\phi)\, M(\phi)\; w^2/4 \tag{4.6.1}$$

where $M(\phi)$ tends to one when ϕ vanishes, to fit with the well-known result for a single particle. We are thus in a very special case where k_f is not obtained from some evolution equation, but is given by an algebraic equation relating it to ϕ and $\mathbf{w}$. We will now consider the consequences of assumption (4.6.1) on the mean momentum equations of the two phases.

When written for two incompressible phases, the two conservation equations for mass can be presented as

$$d_p\phi/dt + \phi\, \nabla.\mathbf{v}^p = 0 \quad \text{and} \quad \nabla.\mathbf{u} = 0 \quad . \tag{4.6.2}$$

When combined with the above expression for k_f, these two equations lead to

$$\rho_f\, d_f k_f/dt = \rho_f\, d_p k_f/dt - \rho_f \mathbf{w}.\nabla k_f$$

$$= (\mathbf{v}_p - \mathbf{u}).[\, \phi\, d_p\mathbf{K}/dt + \rho_f(\partial M/\partial\phi)(\phi^2/4)\, \mathbf{w}\, (\nabla.\mathbf{v}_p) - \rho_f^\circ \nabla k_f - \rho_f^\circ(\phi M(\phi)/4)\, \mathbf{w}\, (\nabla.\mathbf{v}_p)\,]$$

$$= (\mathbf{v}_p - \mathbf{u}).[\, \phi\, d_p\mathbf{K}/dt + \rho_f(\partial M/\partial\phi)(\phi^2/4)\, \mathbf{w}\, (\nabla.\mathbf{v}_p)\,] - \nabla.\, \rho_f k_f \mathbf{w}$$

$$= -\,[\phi\mathbf{K} \otimes (\mathbf{v}_p - \mathbf{u})] : \nabla\mathbf{u} + (\mathbf{v}_p - \mathbf{u}).[\, \phi\, d_p\mathbf{K}/dt + \phi K_i \nabla u_i + \nabla p_{am}\,] - \nabla.(\rho_f^\circ k_f + p_{am})(\mathbf{v}_p - \mathbf{u})$$

where $\mathbf{K}$ is the Kelvin impulse and p_{am} is the *added-mass pressure*. Both $\mathbf{K}$ and p_{am} are related to the fluctuational kinetic energy of the fluid and they are respectively defined as

$$\mathbf{K} = \rho_f^\circ M(\phi)\, (\mathbf{v}_p - \mathbf{u}) / 2 \quad . \tag{4.6.3}$$

and

$$p_{am} = -\, \rho_f^\circ\, (\phi^2/4)\, (v_p - u)^2\, dM/d\phi \, . \tag{4.6.4}$$

Let us detail the last three steps of the above calculation of $\rho_f d_f k_f/dt$. Had we stopped at the second step and compared the result with (4.5.3), we would have concluded that $\sigma^k = 0$, $\mathbf{q}^k = 0$ and $\mathbf{f}^k \neq 0$. But the expression obtained for $\mathbf{f}^k$ would not have been satisfactory because of the presence of forces proportional to ϕ in the dilute limit, and which do not exist in the case of a single particle. Had we stopped at the third step, we would have obtained a much better expression for $\mathbf{f}^k$, but σ^k would have still been null, while its trace should always equal $-2(\rho_f k_f + \rho_p k_p)$ according to definitions (3.3.5) and (3.3.13). So, the fourth step is necessary and comparing with (4.5.3) we finally deduce

$$\sigma^k = -\phi \mathbf{K} \otimes (\mathbf{v}_p - \mathbf{u}) \tag{4.6.5}$$

$$\mathbf{f}^k/(1-\phi) = -\phi\,[d_p\mathbf{K}/dt + K_i \nabla u_i] - \nabla p_{am} \tag{4.6.6}$$

and

$$\mathbf{q}^k = (\rho_f^\circ k_f + p_{am})(\mathbf{v}_p - \mathbf{u}) \quad . \tag{4.6.7}$$

It is noteworthy that $\mathbf{f}^k$ includes the time-derivative of the Kelvin impulse (also called the added-mass force) as well as the added-mass pressure gradient force. Since p_{am} is a function of ϕ and $\mathbf{v}_p - \mathbf{u}$, the related $\nabla\phi$ force involves $\partial p_{am}/\partial\phi$. The stability of the flow is ensured whenever $\partial p_{am}/\partial\phi > 0$ because the $\nabla\phi$ force is then directed towards regions of the flow with a lower ϕ. Unfortunately, it seems that all the calculations made on particles with a frozen configuration ($k_p = 0$) resulted in a function $M(\phi)$ such that $\partial p_{am}/\partial\phi < 0$. Hence, added-mass is a phenomenum which is not favourable to the flow stability.

Since we neglected any dissipative phenomena, the general results (4.5.4) and (4.5.5) become

$$\sigma = \sigma^k - p\,I \quad \text{and} \quad \mathbf{f} = \mathbf{f}^k + \mathbf{f}_\perp$$

The above calculation is unable to give explicit expressions for the pressure p and the lift-force $\mathbf{f}_\perp$. Despite of this shortcoming, it is still interesting in so far as expressions (4.6.5) and (4.6.6) have been proved to rely on assumption (4.6.1) only. In fact, we have met here the first application of the general method suggested by (4.3.8) : The mean features of σ and $\mathbf{f}$ can be deduced from the knowledge of $< \sigma^\circ : \nabla \mathbf{v}^\circ >$ (null in the present case) together with the time rate of change of k_p and k_f (here deduced from (4.6.1)).

4.7 *Pseudo-turbulence*

In the previous section, we assumed that all the particles were moving with the same translational velocity so that k_p was null. Such an assumption is not tenable because of the intrinsic instability of any flow with a vanishing k_p. Let us consider a non-viscous mixture first. If all the particles move with the same velocity, they are certainly located at the apexes of some regular lattice. If one of the particles is slightly perturbed from its equilibrium position, we would like to know if the resulting force exerted by all the other particles will force it to recover its initial position or not. In fact the answer is not and any regular array of particles is unstable whenever it moves relative to the fluid phase. This is in fact suggested by the calculation of $M(\phi)$ for some peculiar lattices which shows that M is an *increasing* function of ϕ in all known cases. Then, the $\nabla\phi$ force present in $\mathbf{f}^k$ is unable to prevent the gathering of particles. One could hope these pessimistic conclusions to be tempered by the existence of some powerful enough viscous damping. Again, this is not the case : due to the indirect interactions between particles produced by the ambient fluid (the hydrodynamic interactions), the existence of a chaotic particle motion is the rule and not the exception.

At variance with Brownian motion, these fluctuations desappear when the flow stops, and contrary to the case of turbulence in a one-phase fluid, they exist however slow is the flow. For these reasons, the case of a k_p produced by hydrodynamic interactions between particles is called pseudo-turbulence. If hydrodynamic interactions between particles will take part in the appearance of k_p, they will also induce some k_f (again for reasons connected with the boundary conditions for the velocity on the interfaces) and the magnitude of k_f will depend on the magnitude of k_p. The problem of pseudo-turbulence can thus be formulated by a double question :

i) How the evolution equation for k_p is coupled to the suspension flow ?

ii) What is the link between k_f and k_p ?

This problem is far from being settled. Just to give a flavour of the kind of results that are needed to develop a complete model of pseudo-turbulence, we give below a very crude answer to the two above questions for the special case of rigid spheres moving in a non-viscous and non-compressible fluid. The two crude answers are

$$\rho^* \, d_p k_p/dt = -\rho^* \mathbf{R}_p : \nabla \mathbf{v}_p - \nabla . \rho^* \mathbf{\Lambda}_p \qquad (4.7.1)$$

and

$$(1-\phi)\, k_f = (\phi\, M(\phi)/4)\ [(\mathbf{v}_p - \mathbf{u})^2 + 2k_p] \qquad (4.7.2)$$

where ρ^*, $\mathbf{R}_p$ and $\mathbf{\Lambda}_p$ are respectively defined as

$$\rho^* = \phi\rho_p{}^\circ + (\phi/2)\, M(\phi)\ \rho_f{}^\circ \qquad (4.7.3)$$

$$\rho_p \mathbf{R}_p = \langle \chi_p \rho^\circ\, \mathbf{v}' \otimes \mathbf{v}' \rangle \qquad (4.7.4)$$

and

$$\rho_p \mathbf{\Lambda}_p = \langle \chi_p \rho^\circ (v'^2/2) \mathbf{v}' \rangle \qquad (4.7.5)$$

The evolution equation (4.7.1) is in fact deduced from (4.1.1) for the special case of spheres moving in a non viscous and potential fluid. The replacement of ρ_p by ρ^* can be considered as a manifestation of added-mass effects in the chaotic motion of the particles. The fluid fluctuational energy is partly due to the mean relative motion and partly to the chaotic particle motion. With (4.7.1) and (4.7.2) it is not so difficult to calculate the sum $\rho_p d_p k_p/dt + \rho_f d_f k_f/dt$ and to present it in a form similar to the right-hand side of (4.5.3). The pitfalls are similar to those already encountered for added-mass in the previous section and the final result is an equation similar to (4.5.3) with

$$\sigma^k = (\mathrm{IV}.6.5) - \rho^* \mathbf{R}_p \qquad (4.7.6)$$

$$\mathbf{f}^k/(1-\phi) = (IV.6.6) - \nabla.\, \rho^* R_p - \nabla p_k \quad (4.7.7)$$

and

$$\mathbf{q}^k = (IV.6.7) + \rho^* \Lambda_p + (\mathbf{v}_p - \mathbf{u}).(\rho^* R_p + p_k I) \quad (4.7.8)$$

where contributions from the added-mass are the same as those written in (4.6.5) to (4.6.7). As to p_k, it can be thought of as the *fluctuational pressure* associated with the chaotic motion and it is defined as

$$p_k = \rho_f^{\circ} (\phi^2/2)\, k_p\, dM/d\phi\,. \quad (4.7.9)$$

After comparing with the added-mass pressure (4.6.4), note that p_k is positive (in all known cases the derivative $dM/d\phi$ is positive) and that it behaves like $\phi^2 k_p$ in the dilute limit.. The tensor $\rho^* R_p + (p_k + p_{am})I$ can be considered as the *effective particle stress.* The flow stability would be ensured with $\partial(p_k + p_{am})/\partial\phi > 0$. Our knowledge of k_p is not sufficient to give a quantitative answer, but it is clear that the particle chaotic motion induces forces that oppose those developped by added-mass effects, hence it is a good candidate to restore the stability of the suspension.

Needless to say that the starting assumptions (4.7.1) and (4.7.2) are questionable and must be improved. In particular, a source term should appear on the right-hand side of (4.7.1), involving for example k_f and (or) the average relative velocity. Moreover, it cannot be taken for granted that the same function $M(\phi)$ relates k_f to the mean relative motion and to the particle chaotic motion (although this holds in the dilute limit). Lastly, $M(\phi)$ corresponds to some spatial distribution of the particles and the average particles configuration is likely to change with the magnitude of k_p, so that it would be more correct to consider (4.7.2) with $M(\phi,k_p)$ Our only objective in this section was to stress on the importance of the fluctuational kinetic energies and their link with the average equations of motion for the two phases. The story of pseudo-turbulence is still in its infancy.

4.8 *Bubbly Fluids*

A gas-liquid mixture will provide us with another example of a suspension for which k_p and k_f are important quantities. Here, the main origin of k_p is the rate of change of the volume of the bubbles, and that motion will also affect the liquid and induce some k_f. To simplify the issue, we will suppose that all the bubbles translate with the same velocity (no pseudo-turbulence) and that we can discard the kinetic effects of the relative motion (no added-mass) as compared to those of the expansion-contraction motion. Last but not least, we suppose that the bubbles do not deform, that they behave like spheres with a variable radius a, and that all spheres in the vicinity of point $\mathbf{x}$ at time t have the same radius $a(\mathbf{x},t)$. This is a considerable amount of assumptions which have transformed our bubbly liquid into some idealized suspension of spheres with variable radius. In fact, we want to present the simplest case for which k_p and k_f are the consequences of the particle internal motion (by internal motion we mean any motion except the translational one).

Besides its role in the fluctuational kinetic energies, the expansion-contraction motion will also act on the internal energy. When expanding or contracting, the bubbles do perform some work. The standard pressure work - p dV will here be transformed into some - $\Delta p\, d\phi$

term where ϕ is the bubble volume fraction and Δp is a pressure difference that will be determined hereafter. That compressional energy will compete with the fluctuational kinetic energy, and we will determine the effect of the variable radius on the suspension stress as well as the evolution equation of that radius in terms of the above pressure difference.

For our spherical particles, da/dt represents the time-derivative of the radius in a frame following the bubble along its translational motion with velocity **v** (note that **v** is here identical with **u** because of the assumption $v_p = v_f$). The internal motion of the bubble can be more or less complicated, but if one assumes that the expansion (or contraction) is homogeneous all over the bubble, then

$$\mathbf{v}_p^\circ = \mathbf{v} + (\mathbf{r}/a)\, da/dt \qquad (4.8.1)$$

where **r** is the position relative to the spheres center. The above velocity field implies $k_p = (3/5)\,(da/dt)^2$, while we expect the fluid fluctuational energy to appear as

$$k_f = (3/2)\,\phi\, Q(\phi)\,(da/dt)^2 \qquad (4.8.2)$$

where $Q(\phi)$ is a function of the particle volume fraction with unit value in the dilute limit (to fit with the well-known velocity field of the fluid around a single expanding sphere). It is left as an exercise to prove that, when applied to a suspension of spheres moving without any relative velocity in a incompressible fluid, the two conservation laws for mass can be expressed as

$$d\phi/dt = \phi\,(1-\phi)\,(3/a)\, da/dt \qquad (4.8.3)$$

and

$$\nabla.\mathbf{u} = \phi\,(3/a)\, da/dt \qquad (4.8.4)$$

In a bubbly liquid, $\rho_p k_p$ is much smaller than $\rho_f k_f$ and we will neglect it. Combining the expression (4.8.2) for k_f with the mass conservation (4.8.3) and (4.8.4), we obtain an explicit expression for the sum $\rho_p d_p k_p/dt + \rho_f d_f k_f/dt$ and the result can be presented as

$$\rho_p\, d_p k_p/dt + \rho_f\, d_f k_f/dt = \sigma^k : \nabla u + \Delta p^k\,(3/a)\, da/dt \qquad (4.8.5)$$

where

$$\sigma^k = -(2/3)\,\rho_f\, k_f\, I \qquad (4.8.6)$$

while the pressure difference Δp^k represents the inertial effects in the fluid when the bubble radius is changing

$$\Delta p^k = \phi\,\rho_f\,[\,Q\, a\, d^2a/dt^2 + (3/2)\,(da/dt)^2\,((1-\phi)(d\phi Q/d\phi) + (2/3)\,\phi Q)\,] \quad . (4.8.7)$$

Another consequence of the velocity field (4.8.1) concerns the rate of compressional work in the particulate phase which can be written as

$$< \chi_p \, p_p^\circ \, \nabla.\mathbf{v}_p^\circ > \; = \; \phi \, p_p \, (3/a) \, da/dt \qquad (4.8.8)$$

where p_p is the average bubble pressure. Because of the absence of relative velocity, the dissipation rate is connected to two processes only, the shear motion of the whole suspension and the rate of change of the bubble radius (not taken into account in (4.5.2) which was written for rigid particles), and Φ will appear as

$$\Phi = \tau^\eta : \nabla\mathbf{u} + \Delta p^\eta (3/a) \, da/dt \qquad (4.8.9)$$

where Δp^η is a pressure difference due to viscous processes in the bubble expansion. The explicit expressions of Δp^η and τ^η are such as to guarantee the positiveness of the dissipation rate. This is the case with (cf. (1.2.7))

$$(\tau^\eta)_{ij} = \eta_f \, \eta(\phi) \, (\partial u_i/\partial x_j + \partial u_j/\partial x_i) \qquad (4.8.10)$$

and

$$\Delta p^\eta = 4 \, \eta_f \, \eta^*(\phi) \, \phi \, (1/a) \, da/dt \qquad (4.8.11)$$

where η_f is the viscosity of the suspending fluid while $\eta(\phi)$ and $\eta^*(\phi)$ are two functions of the particle volume fraction which, according to single-particle results, tend to one in the dilute limit. With the results (4.8.8) and (4.8.9), and taking the liquid incompressibility into account, one gets from (4.4.2)

$$< \sigma^\circ : \nabla\mathbf{v}^\circ > \; = \; (\tau^\eta - pI) : \nabla\mathbf{u} + (\Delta p^\eta - \phi(p_p - p)) \, (3/a) da/dt \qquad (4.8.12)$$

where p is the average suspension pressure. We now introduce (4.8.5) and (4.8.12) into the left-hand side of equation (4.3.8), and a term to term identification leads to

$$\mathbf{q}^k = 0 \qquad (4.8.13)$$

$$\sigma = \tau^\eta + \sigma^k - p\,I \qquad (4.8.14)$$

and

$$0 = \Delta p^\eta + \Delta p^k - \phi(p_p - p) \qquad (4.8.15)$$

With results (4.8.7) and (4.8.11), the last equality can be transformed into an equation for the evolution of the bubble radius

$$\rho_f Q(\phi) a \, d^2a/dt^2 + (3/2)\rho_f (da/dt)^2 \, [(1-\phi)(d\phi Q/d\phi) + (2/3)\,\phi Q] + 4 \, \eta_f \, \eta^*(\phi)(1/a) da/dt = p_p - p$$

That equation is nothing but the extension to non-dilute bubbly liquids of the Rayleigh-Plesset equation for a single bubble. To sum up, our idealized suspension of spheres is described by the momentum equation (3.3.11) where Π is replaced by the above σ, and by an extension of the Rayleigh-Plesset equation describing the evolution of the new dynamic variable a(x,t). Needless to say that the above description is oversimplified and that taking both the relative motion and pseudo-turbulence into account is necessary to describe real bubbly liquids, not to mention the role of the variance in the bubble radius and of the bubble deformation.

4.9 *Results including phase transitions*

When the mass transfers between phases are taken into account, the main modifications concern the exchanges at the interfaces. The evolution equation for ε_n and k_n can still be written as in (4.1.2) and (4.2.1), provided the energy exchanges are redefined as

$$E_n^k = <\mathbf{v}'_n.\sigma_n^\circ.\mathbf{n}_n\,\delta_I> + <\rho_n^\circ(v_n'^2/2)(\mathbf{v}_I - \mathbf{v}_n^\circ).\mathbf{n}_n\,\delta_I> - M_n k_n \qquad (4.9.1)$$

and

$$E_n^T = - <\mathbf{q}_n^\circ.\mathbf{n}_n\,\delta_I> + <\rho_n^\circ\varepsilon_n^\circ(\mathbf{v}_I - \mathbf{v}_n^\circ).\mathbf{n}_n\,\delta_I> - M_n\varepsilon_n \qquad (4.9.2)$$

As a consequence, results (4.2.6) and (4.3.1) are modified into

$$E_n^k + E_n^T = E_n^* - M_n(\varepsilon_n + k_n) \ . \qquad (4.9.3)$$

and

$$E_p^k + E_f^k = -(\mathbf{v}_p-\mathbf{v}_f).(\mathbf{P}_p - M_p(\mathbf{v}_p+\mathbf{v}_f)/2) - \sum_n M_n k_n +$$
$$+ \sum_n <(\mathbf{v}_n^\circ-\mathbf{v}_I).\sigma_n^\circ.\mathbf{n}_n\,\delta_I> + \sum_n <\rho_n^\circ(\mathbf{v}_n^\circ-\mathbf{v}_I)^2/2)(\mathbf{v}_I - \mathbf{v}_n^\circ).\mathbf{n}_n\,\delta_I> \ . \qquad (4.9.4)$$

Instead of (4.3.6) and (4.3.7), the momentum equations are advantageously rewritten as

$$\rho_p\,(d_p\mathbf{v}_p/dt - \mathbf{g}_p) = \phi\,\nabla.\sigma + \mathbf{f} + M_p\,(\mathbf{v}_f - \mathbf{v}_p)/2 \qquad (4.9.5)$$

$$\rho_f\,(d_f\mathbf{v}_f/dt - \mathbf{g}_f) = (1-\phi)\,\nabla.\sigma - \mathbf{f} + M_p\,(\mathbf{v}_f - \mathbf{v}_p)/2 \qquad (4.9.6)$$

where the former definition (4.3.4) for **f** is modified into

$$\mathbf{f} = \mathbf{P}_p - M_p(\mathbf{v}_p+\mathbf{v}_f)/2 + \nabla.\sigma_p - \phi\,\nabla.\sigma \qquad (4.9.7)$$

Concerning the evolution of the total fluctuational kinetic energy and total internal energy, it is transformed from (4.3.3) and (4.4.1) into

$$\sum_n (\rho_n\,d_n k_n/dt + M_n k_n) = \sigma:\nabla\mathbf{u} - \mathbf{w}.\mathbf{f} - \rho A - \nabla.\mathbf{q}^k + \rho G^* \qquad (4.9.8)$$

and

$$\sum_n (\rho_n\,d_n\varepsilon_n/dt + M_n\varepsilon_n) = \rho A - \nabla.(\mathbf{q}_p^T + \mathbf{q}_f^T) \ . \qquad (4.9.9)$$

The coupling energy ρA has now replaced $<\sigma^\circ:\nabla\mathbf{v}^\circ>$ and is defined as

$$\rho A = <\sigma^\circ:\nabla\mathbf{v}^\circ> - \sum_n <(\mathbf{q}_n^\circ - \rho_n^\circ\varepsilon_n^\circ(\mathbf{v}_I - \mathbf{v}_n^\circ)).\mathbf{n}_n\,\delta_I>$$

$$= < \sigma^\circ : \nabla \mathbf{v}^\circ > \; - \; \Sigma < (\mathbf{v}_n^\circ - \mathbf{v}_I) . \sigma_n^\circ . \mathbf{n}_n \, \delta_I > \; - \; \Sigma < \rho_n^\circ (\mathrm{v}_n^\circ - \mathrm{v}_I)^2 / 2)(\mathbf{v}_I - \mathbf{v}_n^\circ) . \mathbf{n}_n \, \delta_I >$$

$$= < \sigma^\circ : \nabla \mathbf{v}^\circ > \; + \; < (\mathbf{v}_p^\circ - \mathbf{v}_f^\circ) . \sigma_f^\circ . \mathbf{n}_f \, \delta_I > \; + \; < ((\mathrm{v}_p^\circ - \mathrm{v}_f^\circ)^2 / 2) \, \rho_p^\circ (\mathbf{v}_I - \mathbf{v}_p^\circ) . \mathbf{n}_p \, \delta_I > ,$$

the second and third expressions being deduced from the boundary conditions (3.2.7), (3.3.8) and (3.5.9).

REFERENCES

1. Hinch, E.J.: An averaged equation approach to particles interactions in a fluid suspension, J. Fluid Mech. 83 (1977), 695-720.

2. Batchelor, G.K.: The stress-tensor in a suspension of force-free particles, J.Fluid Mech. 41 (1970), 545-570.

3. Shaqfeh, E. and G. Fredrickson : The hydrodynamic stress in a suspension of rods, Phys. Fluids A2 (1990), 7-24.

4. Leal, L.G.: Macroscopic transport properties of a sheared suspension, J. Coll. Interface Sc. 58 (1977), 296-311.

5. Buyevich, Y.A.: Heat and mass transfer in disperse media I. Average field equations II. Constitutive equations, Int. J. Heat Mass Transfer 35 (1992), 2445-2463.

6. De Groot, S.R. and P. Mazur : Non-equilibrium Thermodynamics (ch.III), North-Holland, Amsterdam, 1969.

7. Maugin, G.A. and W. Muschik : Thermodynamics with internal variables, I. General concepts II. Applications, J. Non-Equilib. Thermo. 19 (1994), 217-289.

8. Lhuillier, D. : From molecular mixtures to suspensions of particles, J. Phys.II France 5 (1995), 19-36.

9. Barthes-Biesel, D. and A. Acrivos : Deformation and burst of a liquid droplet freely suspended in a linear shear field, J. Fluid Mech., 61(1973), 1-21.

10. Hand, G.L. : A theory of anisotropic fluids, J. Fluid Mech. 13 (1962), 33-46.

11. Onuki, A.: Dynamic equations of polymers with deformations in the semi-dilute regions, J. Phys.Soc. Japan, 59 (1990), 3423-3426.

12. Doi, M.: Effects of viscoelasticity on polymer diffusion, in: Dynamics and Patterns in Complex fluids (Ed. A. Onuki and K. Kawasaki) Springer Proceedings in Physics (1990) vol.52.

13. Drew, D.A. : Mathematical modelling of two-phase flows, Ann.Rev.Fluid Mech. 15 (1983), 261-291.

14. Lhuillier, D.: Ensemble averaging in slightly non-uniform suspensions, Eur.J.Mech., B/Fluids, 11 (1992), 649-661.

15. Zhang, D.Z. and A. Prosperetti : Averaged equations for inviscid disperse two-phase flow, J.Fluid Mech. 267 (1994), 185-219

16. Nigmatulin, R.I. : Dynamics of Multiphase media , Hemisphere,New-York, 1990.

SUSPENSIONS OF CAPSULES

D. Barthes-Biesel
Compiègne University of Technology, U.R.A. C.N.R.S. 858,
Compiègne, France

ABSTRACT

The motion of capsules freely suspended into another liquid subjected to flow is studied. Experimental evidence regarding the deformation of artificial capsules or of red blood cells in shear flows is presented. Results of filtration experiments conducted on red cell suspensions are also discussed. The equations describing the mechanics of a single capsule are presented. Perturbation solutions obtained for initially spherical particles are discussed together with approximate solutions for ellipsoidal capsules. In the case of non spherical particles subjected to large deformations, numerical models must be devised, and some new results are presented.

1. INTRODUCTION

A capsule consists of an internal medium (pure or complex liquid, granular solid, ...), enclosed by a deformable membrane that is usually semi-permeable. Capsules are frequently met in nature (most living cells may be considered as capsules) or in industrial processes. As a consequence, encapsulation is presently a very active field of research, with applications in medicine, pharmaceutics and industry. Capsules are found in a variety of sizes from a few microns (liposomes, living cells) to a few millimeters (artificial capsules). Furthermore, they may take different rest shapes (e.g., discoïdal geometry of a red blood cell), and the mechanical properties of the internal medium and of the membrane may vary widely. The main problems presently met with encapsulation deal with fabrication processes: how to produce capsules with a given size distribution and proper transport properties; how to prevent undesirable chemical reactions between the membrane, the internal contents and/or the external medium. However, it is also important to assess the mechanical properties of the capsules, to predict their deformation under stress and the occurrence of breakup. This is a non trivial problem since capsule deformability depends on three types of independent intrinsic physical properties:

- initial geometry;
- internal liquid rheological properties;
- interface constitutive behaviour.

Here, we shall deal with capsules filled with a homogeneous viscous liquid and designed to be suspended in another viscous liquid subjected to flow. A typical example is the red blood cell, which is suspended in plasma. In its normal rest state, the red blood cell is a small biconcave disc (diameter of order 8μm), filled with an hemoglobin solution (viscosity 8 to 10 mPa s) and enclosed by a lipid bilayer lined by a protein network. The study of the mechanical properties of red blood cells has been actively going on for the past 40 years. A number of experimental devices have been proposed to measure cell deformability and some theoretical models have been developed to interpret the data (see the reviews of Fung [1] and of Skalak et al. [2]). However, red blood cells are difficult to measure owing to their small size and to their sensitivity to external physicochemical conditions. Furthermore, they are also difficult to model because they are flaccid, have an anisotropic initial configuration and are usually subjected to large deformations. On the other hand, the study of artificial capsules, large enough to be easily observed and with controlled mechanical properties of the membrane and of the internal medium is very useful for understanding the main physical phenomena that prevail when such particles are subjected to flow.

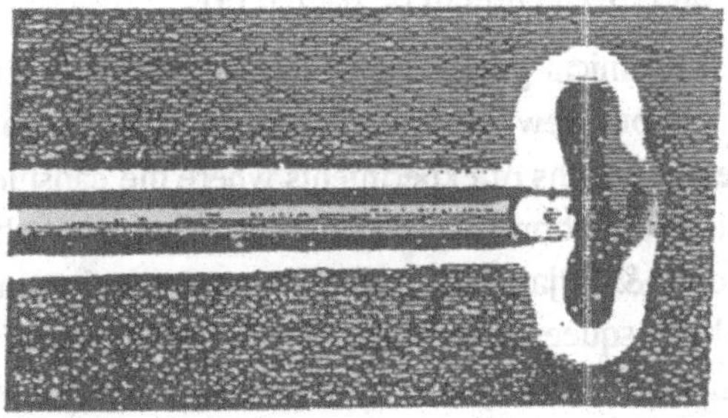

Figure 2.1 Suction of a normal red blood cell in a micropipet. (picture obtained by C. Bucherer, Unité de Biorhéologie, Hôpital Pitié-Salpêtrière)

It is the objective of this paper to present some of experimental evidence that is available on blood cells and artificial capsules (section **2**). The generally agreed upon formulation of the motion and deformation of a capsule freely suspended in an unbounded shear field is discussed in section **3**, together with some solutions of the equations of motion (section **4**). The flow of capsules through small pores is studied in section **5**.

2. EXPERIMENTAL OBSERVATIONS OF CAPSULE DEFORMATION.

Experimental studies of capsule deformability usually strive to determine independently the membrane mechanical properties and the overall deformation and behaviour of a capsule in a shear flow.

2.1. Determination of membrane properties.

In the case of small capsules (with a typical size scale of the order of 10µm or less), research has been mostly conducted on biological cells, and specifically on blood cells. The thickness of the bilayer is of order 10^{-8} m, so that the membrane can be treated as a two-dimensional elastic solid. The mechanical properties are measured with a micropipet with a 1 to 2 µm internal diameter. A small part of the membrane is sucked into the pipette (Figure 2.1), and the height of aspiration is measured as a function of the applied depression. Such measurements, as well as relaxation experiments after suction, are used to evaluate the membrane shear elastic modulus and surface viscosity. Micropipet experiments performed on red blood cells indicate that the membrane is easily sheared, strongly resists any increase in the local surface area, has a small surface bending modulus and a shear-thinning surface viscosity. Details on red blood cell membrane mechanics can be found in two review papers by Hochmuth & Berk [3] and by Hochmuth & Waugh [4].

The measurement of the mechanical properties of large capsules is a relatively new field of research, and there are yet but few published results. The overall average membrane properties may be assessed by means of experiments where the capsule is squeezed between two plates and its overall deformation is recorded as a function of the squeezing force. A theory developed by Lardner & Pujara [5] enables to compute the surface elastic modulus, once the deformation and the squeezing forces are measured. Such a procedure has been used to measure the membrane elasticity of sea-urchin eggs (Hiramoto [6]) or of artificial capsules enclosed by a nylon membrane (Chang & Olbricht [7],[8]).

Another method consists in creating a flat film of the membrane material and in measuring directly the mechanical properties of this film by means of a surface torsion rheometer (Burger & Rehage [9]). When feasible, this technique is quite powerful since it gives access to eventual complicated material properties of the membrane such as non-linear stress-strain laws and visco-elasticity.

2.2. Motion of a capsule in shear flow.

The common principle underlying most of the experiments is to subject capsules to viscous stresses exerted by the flow of the suspending medium and to observe the overall deformation of a single particle. There are many available experimental results on the motion of liquid droplets (for emulsion technology purposes mainly), of blood cells (for physiological and medical applications), and of some artificial capsules. In all cases, the objective is to measure the deformation of a capsule with characteristic dimension A (e.g., the radius of the sphere that has the same volume), filled with a viscous incompressible liquid of viscosity $\lambda\mu$, enclosed in a deformable interface with a surface elastic modulus E_S, and suspended in another viscous liquid with viscosity μ. Different shear flows are used, defined by the velocity field:

$$\mathbf{u}_\infty = G\,[\,\mathbf{E}(t) + \mathbf{\Omega}\,(t)].\,\mathbf{x}, \tag{2.1}$$

where G is the magnitude of the shear rate and where **E** and Ω denote respectively the rate of strain and the vorticity tensors that depend only on time. In the gap of a Couette or of a cone-and-plate viscometer, a simple shear flow exists such that

$$E_{12} = E_{21} = \Omega_{12} = -\,\Omega_{21} = 1/2. \tag{2.2}$$

where x_1 x_2 is the shear plane The four-roller flow cell designed by Bentley & Leal (1986) produces a pure straining motion in the x_1 x_2 plane with no vorticity:

$$E_{11} = - E_{22} = 1. \tag{2.3}$$

Another flow of interest is the elongational motion, encountered experimentally in extrusion processes, and very popular with theoreticians because it is axisymmetric. In the case of an elongation along the x_1 direction, the flow field is given by:

$$E_{11} = - 2 E_{22} = - 2 E_{33} = 2. \tag{2.4}$$

In (2.1) to (2.4) all the other components of E and of Ω are zero.

The capsule deformation is usually measured by observing the projection of its profile on the principal shear plane. The deformation D is then defined as:

$$D = (L - B) / (L + B), \tag{2.5}$$

where L and B denote respectively the length and the breadth of the deformed particle profile. A simple dimensional analysis indicates that the capsule motion and deformation depend on two main dimensionless parameters:

- the viscosity ratio λ between the internal and external liquids,
- the Capillary number, which measures the ratio of viscous to elastic forces:

$$C = \mu G A / E_S, \tag{2.6}$$

Other parameters must obviously be introduced to define the type of flow (e.g., the ratio of vorticity to strain), the particle initial geometry and different membrane properties (visco-elasticity, area incompressibility, bending resistance, ...).

To get some insight on capsule mechanics, the behaviour of an artificial capsule suspended in a shear field is presently studied with simple particles having a Newtonian internal liquid, a presumably isotropic membrane and an initially spherical geometry. The principle consists is observing the deformation of the particle in a hyperbolic field created in a four roller flow cell (Chang & Olbricht [7], Akchiche [10], Barthes-Biesel [11]) or in a simple shear field created between counter-rotating flat plates (Burger & Rehage [9]) or cylinders (Chang & Olbricht [8]). Similar observations have also been performed on biological cells suspended in simple shear or elongational flows.

i) Capsule in pure straining flow.

In the case of a hyperbolic flow Akchiche [10] has compared the behaviour of liquid drops and of capsules (diameter of order 3mm), enclosed by a polymerized polylysine membrane, coated by an alginate film and filled with an aqueous 1% alginate solution. The external suspending liquids are viscous silicone oils. The viscosity ratio in both systems is small ($\lambda<0.02$). The liquid drop is very elongated, eventually exhibits pointed ends and can continue to elongate without bursting. This is in accordance with observations by Bentley & Leal [12]. By contrast, the capsule deformation is much smaller since it is limited by the membrane extensibility. For values of C in excess of 0.02, the capsule bursts by developing pointed ends where the membrane is ruptured and through which small droplets of the internal liquid are expelled. The difference between the two deformation versus Capillary number curves (Figure 2.2) shows clearly the shielding role of the solid deformable interface. Chang & Olbricht [7] have completed similar experiments in a hyperbolic flow, on capsules with a nylon membrane in a cross-linked rubbery state. Their results (as estimated from Figure 6 of [7]) have also been plotted on Figure 2.2. The nylon capsule exhibits a quite different behaviour from the polylysine capsule, since its deformation reaches a limiting value for high shear. This indicates some shear hardening, a common phenomenon in crosslinked polymers undergoing extension. Furthermore, after cessation of flow, the capsule does not return to its initial spherical state but keeps a small residual deformation.

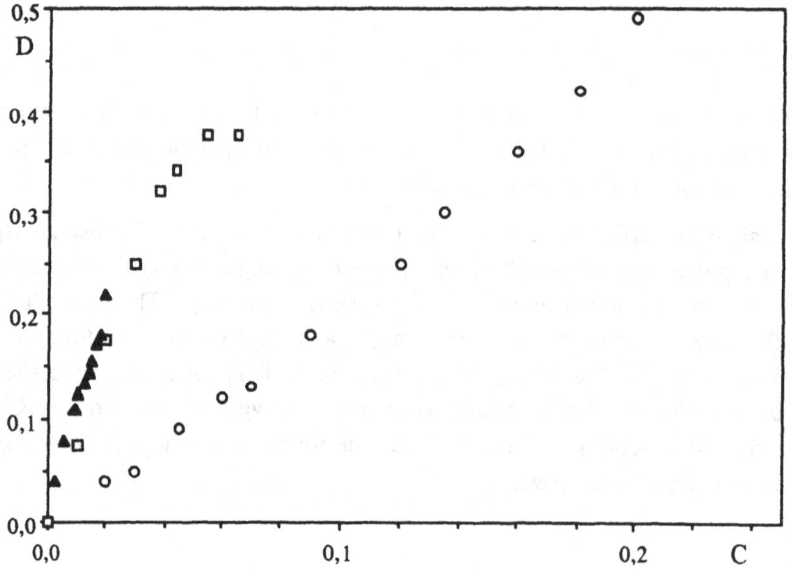

Figure 2.2 Pure straining flow: deformation of a liquid drop (o) and of a capsule with a polylysine membrane (▲). The results for the nylon membrane (□) are obtained from [7].

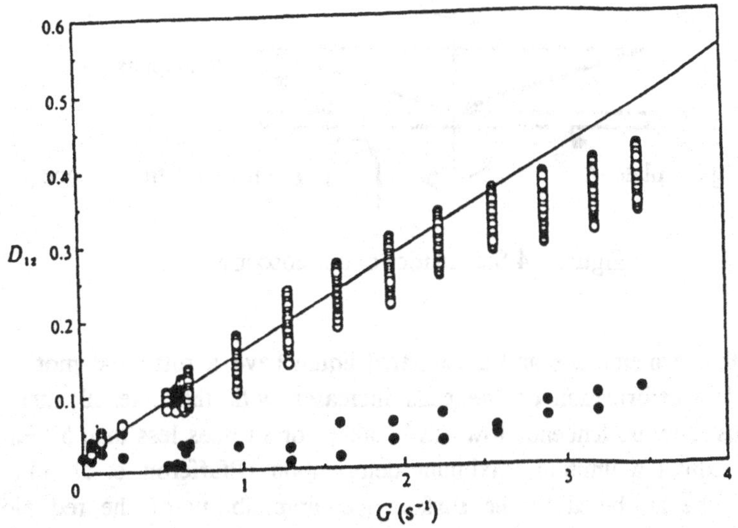

Figure 2.3 Deformation of a capsule in simple shear flow. The capsule deformation oscillates about a mean value. After cessation of flow a permanent deformation remains (filled symbols). From [8].

ii) Capsule in simple shear flow.

Chang & Olbricht [8] have also observed the same nylon capsules in a simple shear flow. They measure the deformation (denoted D_{12}) and orientation in the x_1x_2 shear plane as well as in the orthogonal x_2x_3 plane. They find that the capsule deforms and orients with respect to streamlines. For a given constant shear rate, the deformation and orientation are not steady, but oscillate about a mean constant value (Figure 2.3). There is also a small permanent deformation after flow cessation. Furthermore, when the deformation becomes large enough, the capsule breaks. This is attributed to a local thinning of the membrane.

Considerable insight regarding red blood cell mechanics has been gained with a rheoscope, a device first proposed by Schmid-Schönbein & Wells [13]. The rheoscope is a glass counter-rotating cone-and-plate viscometer, in the gap of which is placed a dilute red blood cell suspension. The deformation of the cells that have their center of mass in the zero velocity cone and which are thus stationary, is observed and measured by means of an inverted microscope (Figure 2.4). Then, under the influence of the viscous forces due to the flow, all cells take similar prolate ellipsoidal shapes oriented with respect to the streamlines.

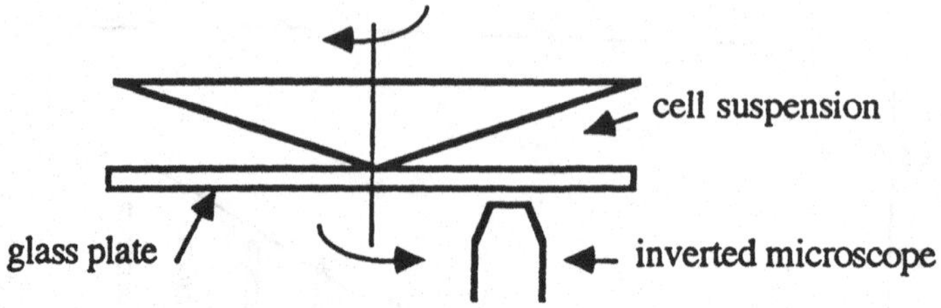

Figure 2.4 Schematics of a rheoscope.

Furthermore, both the membrane and the internal liquid have a rotational motion, called 'tank-treading'. The deformation of the cells increases with the external shear stress. However, the cells show no tendency towards breakup for stresses less than 50 Pa (Figure 2.5), but rather exhibit a limiting maximum deformation (Pfafferott et al. [14]). This phenomenon may be attributed to the surface incompressibility of the red blood cell membrane, which limits the local deformations. When the viscosity ratio λ between the internal and external liquids, is less than about 2, the deformation at a given shear stress level, decreases when λ increases. For larger values of λ, the red cells show a tendency to behave as solid flexible discs and flip rather than tank-tread.

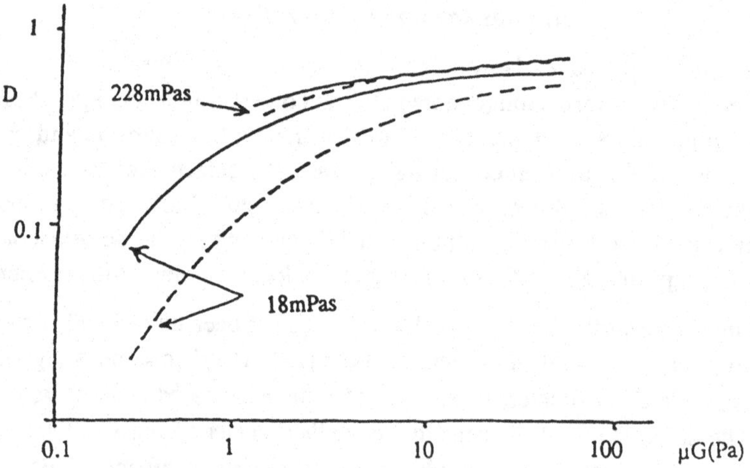

Figure 2.5 Deformation of red blood cells as a function of shear stress, for different suspending medium viscosity: From [14].

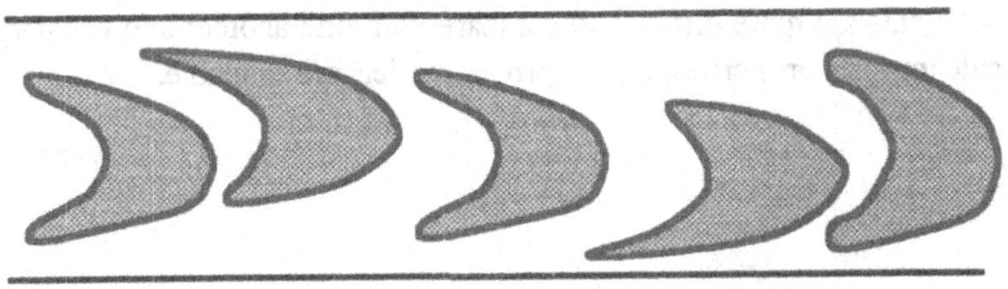

Figure 2.6 Deformation of red blood cells in a capillary vessel.

iii) Capsule in tube flow.

The case of capsules flowing in tubes is also important from the physiological (microcirculation) and engineering (filtration) point of view. Most of the available data deals with red blood cells. Experimental observations of the microcirculation *in vivo* show that the 8μm discoidal cells are able to squeeze into capillary vessels with an internal diameter as small as 4 μm. In order to do so, the cell deforms and takes a parachute or slipper like shape (Figure 2.6). This process is facilitated by the fact that red blood cells are flaccid particles surrounded by an excess area of membrane as compared to the enclosed volume.

Red blood cell deformability is also often measured by means of filtration apparatus. The common underlying principle of the different setups is to force a cell suspension through a filter with pores having a diameter smaller than the cells', and to measure simultaneously the flow rate and the pressure drop across the filter. The experimental protocols differ in the nature of the sieve, the pore geometry, the suspension hematocrits (from 1% to 90%) and the flow driving force (see the review by Nash [15]). With a large number of pores, the measured bulk quantities correspond to averages of the pore level micromechanics. In some other devices with a small number of pores (of order 20 to 30), it is possible to follow the flow of one cell through one pore, and then to take the average of many measurements (Fischer et al. [16]). It is difficult to compare experimental data obtained in different devices, because the measured values depend not only on the imposed flow forces (suspending medium viscosity, tube size, etc...) but also on the intrinsic physical properties of the cells. However, careful experiments (Kiesewetter et al. [17], Schmid-Schönbein & Gaehtgens [18], Drochon et al. [19]) indicate that filtration results are sensitive to the size ratio between the cells and the tube, to the internal viscosity of the cells, but not very sensitive to the properties of the cell membrane.

All those experimental results show clearly that the membrane mechanical properties, the capsule initial shape and the type of flow all have a strong influence on the overall capsule motion. The experiments are quite difficult to compare with one another and to interpret in terms of the capsule intrinsic properties, unless proper models are available.

3. MEMBRANE MECHANICS.

The membrane is usually modeled as a two-dimensional shell with negligible thickness, subjected to large deformations. Bending effects are neglected, and only shear deformations in the membrane plane are considered. As shown by Barthes-Biesel and Rallison [20], the theory of large deformations of membranes can be conveniently expressed with Cartesian tensors. This description is particularly appropriate for problems that involve a strong coupling between fluid and solid mechanics.

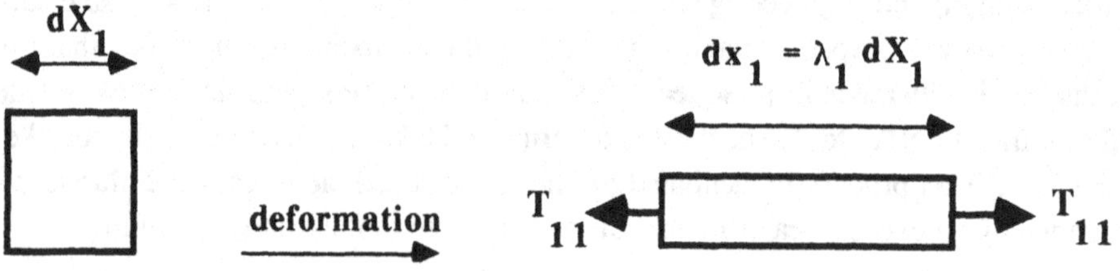

Figure 3.1 Schematics of the deformation of a membrane element.

The deformation of the interface is defined by comparing the position **X** of a membrane material point in the stress-free reference configuration to its instantaneous position **x** (**X**, t) at time t. The outer unit normal to the surface is denoted **N** before deformation and **n** after deformation. A Cauchy-Green deformation tensor $\mathbf{C}_e$ is introduced, with components only in the membrane plane:

$$\mathbf{C}_e = {}^T\mathbf{F}.\ \mathbf{F} - \mathbf{I}, \tag{3.1}$$

where the surface displacement gradient **F** is defined by:

$$\mathbf{F} = (\mathbf{I} - \mathbf{N}.{}^T\mathbf{N}).\partial\mathbf{x}/\partial\mathbf{X}.\ (\mathbf{I} - \mathbf{n}.{}^T\mathbf{n}), \tag{3.2}$$

where **I** is the identity matrix. The eigenvalues of $\mathbf{C}_e$ are simply the principal extension ratios λ_1, λ_2 in the membrane plane. Thus if (Ox_1, Ox_2) are the two principal directions of deformation and stress in the deformed membrane plane, corresponding respectively to directions (OX_1, OX_2) before deformation, then λ_1, λ_2 are defined by (Figure 3.1):

$$\lambda_1 = dx_1 / dX_1, \qquad \lambda_2 = dx_2 / dX_2 . \tag{3.3}$$

Stresses are replaced by tensions, i.e., forces per unit length of deformed membrane. Neglecting inertia, the shell equilibrium equations become (high frequency vibrations of the surface are *a priori* excluded):

$$\mathbf{q} + (\mathbf{I} - \mathbf{n}.^T\mathbf{n}).\nabla.\mathbf{T} = 0, \tag{3.4}$$

where $\mathbf{T}$ denotes the Cauchy tension tensor with components only in the membrane plane, and where $\mathbf{q}$ is the load per unit area exerted by the membrane on the fluids.

The tensor $\mathbf{T}$ is then related to the local deformations of the membrane by means of a rheological constitutive equation. A few classical laws are available to give a phenomenological description of typical mechanical behaviours. Some constitutive equations are presented below, but a more detailed discussion may be found in Barthes-Biesel [11]. The expression for the principal component T_{11} is given explicitly and the corresponding expression for T_{22} can be simply deduced by inverting the roles of the indices 1 and 2. The modulus E_s represents a surface Young modulus. The corresponding value of the surface shear modulus G_s is also given.

i) Linear elasticity.

This law corresponds to Hooke's law adapted to a two-dimensional continuum. It is restricted to small deformations of the interface.

$$T_{11} = E_s\,[\lambda_1^2 - 1 + \nu^2(\lambda_2^2 - 1)] / 2(1 - \nu^2), \tag{3.5}$$

$$G_s = E_s / 2(1+\nu).$$

The surface Poisson ratio ν is unity in case of surface area incompressibility.

ii) Rubber elasticity.

The membrane is treated as an infinitely thin layer of a homogeneous, isotropic, three dimensional incompressible elastomer obeying a Mooney-Rivlin type of law, with surface shear elastic modulus E_s:

$$T_{11} = \frac{E_s}{3\lambda_1\lambda_2}\left\{\Psi\,[\lambda_1^2 - \frac{1}{(\lambda_1\lambda_2)^2}] + \Psi'[(\lambda_1\lambda_2)^2 - \frac{1}{\lambda_1^2}]\right\}, \tag{3.6}$$

$G_s = E_s/3.$

The coefficients Ψ and Ψ' are dimensionless and such that $\Psi + \Psi' = 1$. By allowing Ψ and Ψ' to be functions of the strain invariants, it is possible to model complex elastic behaviours such as strain hardening. The case $\Psi'=0$ corresponds to a neo-Hookean solid.

iii) Red blood cell elasticity.

Skalak et al. [21] have proposed a constitutive law for the red blood cell membrane that takes into account the shearing ability and the large resistance to surface area changes. Two surface elastic moduli are thus introduced: E_s corresponding to shear deformations and A_s corresponding to area changes and such that $A_s/E_s >> 1$ (A_s is of order $10^5 E_s$):

$$T_{11} = E_s \frac{\lambda_1}{4\lambda_2}(\lambda_1^2 - 1) + A_s \lambda_1\lambda_2 [(\lambda_1\lambda_2)^2 - 1]. \tag{3.7}$$

For interpretation of experimental measurements, it is usually enough to use a simplified version of this law, proposed by Evans [22], where the shear behaviour is simply quadratic in λ_1 and where the membrane is assumed to be exactly area incompressible:

$$T_{11} = \frac{E_s}{4}(\lambda_1^2 - 1) + T_0, \tag{3.8}$$

Then T_0 plays the role of an isotropic pressure that must be determined from the additional constraint that expresses the invariance of the local surface area, i.e., $\lambda_1\lambda_2 = 1$. In (3.7) and (3.8) the shear elastic modulus is given by:

$G_s = E_s/4.$

iv) Visco-elasticity.

Visco-elastic effects are usually modeled by means of a linear law where the expansion viscosity is neglected. Thus to any of the above elastic laws, a viscous contribution may be added:

$$T_{11}^{\nu} = 2\mu_s \frac{\partial\lambda_1/\partial t}{\lambda_1} \tag{3.9}$$

where $\partial\lambda_1/\partial t$ denotes a time derivative, and where μ_s is the surface viscosity.

4. MOTION OF A CAPSULE IN A SHEAR FLOW.

In most models the following starting hypothesis are made. The capsule initial unstressed geometry is given. The internal liquid filling the capsule is assumed to be Newtonian and incompressible with a viscosity $\lambda\mu$. The external suspending liquid is also Newtonian and incompressible with a viscosity μ. The membrane is very thin and treated as a two-dimensional interface (this hypothesis has not been used by Brunn [23] who specifically considered a thick shell capsule). It is impermeable to both the internal and external liquids. Buoyancy effects are ignored and the particle Reynolds number of the flow is assumed to be very small.

4.1. General problem statement.

The equations of motion are written in a reference frame (O, x_1, x_2, x_3) centered on the particle center of mass O and moving with it (Figure 4.1). The external liquid is subjected far from the particle to an undisturbed flow field $\mathbf{u}_\infty$, p_∞. All quantities are non dimensionalized: lengths by A, velocities by GA or by $|\mathbf{u}_\infty|$, time by G^{-1} viscous stresses by μG and elastic tensions by E_s. Starred quantities refer to the internal liquid. Under the influence of the viscous stresses, the cell deforms and the equation of its interface is given by:

$$r = (x_1^2 + x_2^2 + x_3^2)^{1/2} = f(x_1, x_2, x_3)\,. \tag{4.1}$$

The function f is itself unknown and must be determined as part of the problem solution. In absence of inertia effects, the internal and external velocity and stress fields, respectively $\mathbf{u}^*$, $\mathbf{u}$, σ^* and σ, are governed by the Stokes equations:

$$\nabla.\mathbf{u} = 0, \qquad \nabla.\sigma = 0, \qquad \text{for } r \geq f, \tag{4.2}$$

$$\nabla.\mathbf{u}^* = 0, \qquad \nabla.\sigma^* = 0, \qquad \text{for } r \leq f. \tag{4.3}$$

For Newtonian liquids, the stresses are given by:

$$\sigma = -p\mathbf{I} + (\nabla\mathbf{u} + {}^T\nabla\mathbf{u}), \tag{4.4}$$

$$\sigma^* = -p^*\mathbf{I} + \lambda\,(\nabla\mathbf{u}^* + {}^T\nabla\mathbf{u}^*), \tag{4.5}$$

where p and p^* denote respectively the external and internal pressure. The associated boundary conditions are:

i) no flow disturbance far from the cell:

$$\mathbf{u} \rightarrow \mathbf{u}_\infty \qquad \text{as} \qquad r \rightarrow \infty, \tag{4.6}$$

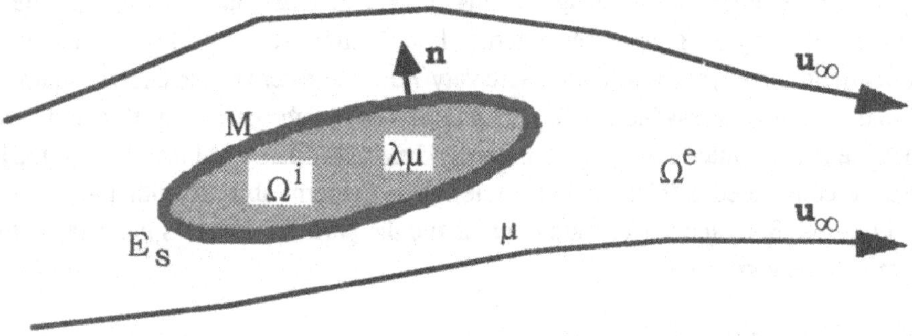

Figure 4.1 Schematics of a capsule freely suspended in a shear flow.

ii) continuity of velocities at the interface:

$$\mathbf{u}^* = \mathbf{u} = \mathbf{u}_s = \partial \mathbf{x} / \partial t \qquad \text{at} \quad r = f, \tag{4.7}$$

where $\mathbf{u}_s$ represents the velocity of the membrane, as measured by the time derivative of the position $\mathbf{x}$ of the interface material points.

iii) dynamic equilibrium of the interface:

$$(\sigma - \sigma^*).\ \mathbf{n} + C\,\mathbf{q} = 0 \qquad \text{at} \quad r = f, \tag{4.8}$$

Condition (4.8) states that the viscous loads exerted by the flow of the internal and external liquids are balanced by the elastic force $\mathbf{q}$ exerted by the deformed membrane. To close the formulation of the problem there remains to relate $\mathbf{q}$ to the surface deformation as shown in section 3.

The solution of equations (4.1) to (4.8), with any constitutive equation (3.5) to (3.9) represents a very difficult problem of continuum mechanics. It is of the free surface type and requires the solution of two flow problems for the internal and external liquids. Furthermore, the treatment of the interface and associated boundary conditions leads to difficulties linked to the following aspects:

- The initial geometry is not necessarily spherical, and thus not isotropic. Furthermore, it may also be prestressed.

- The interface mechanics require the use of the theory of large deformations of shells.

• The interface deformation is usually described with Lagrangian variables, whereas the fluid mechanics use Eulerian variables.

Analytical models have been obtained that are valid for initially spherical capsules subjected to moderate deformations. The motion of a deformed ellipsoidal capsule has also been determined. When other initial shapes and large deformations need to be considered, it is necessary to resort to a numerical solution of this set of equations. Then, it is convenient to recast the Stokes equations (4.2) and (4.3) in integral form. This approach was first taken by Rallison & Acrivos [24] to determine the deformation of a liquid drop in a pure straining motion. The external liquid domain Ω^e is bounded by an outer boundary $\partial\Omega$, that may be solid, fluid or both. The internal liquid is bounded by the deformed membrane, the surface of which is denoted M. The velocity of any point of Ω^e can then be expressed as:

$$\alpha \mathbf{u}(\mathbf{x}) = (1-\lambda)\int_M \mathbf{K}(\mathbf{x}-\mathbf{y}).\mathbf{u}_s(\mathbf{y}).\mathbf{n}(\mathbf{y})dS(\mathbf{y}) - \int_{\partial\Omega} \mathbf{K}(\mathbf{x}-\mathbf{y}).\mathbf{u}(\mathbf{y}).\mathbf{n}(\mathbf{y})dS(\mathbf{y})$$

$$+ \frac{1}{8\pi\varepsilon}\int_M \mathbf{J}(\mathbf{x}-\mathbf{y}).\mathbf{q}(\mathbf{y})dS(\mathbf{y}) + \frac{1}{8\pi}\int_{\partial\Omega} \mathbf{J}(\mathbf{x}-\mathbf{y}).\mathbf{F}(\mathbf{y})dS(\mathbf{y}), \qquad (4.9)$$

where $\alpha=(\lambda+1)/2$ for $\mathbf{x} \in M$, $\alpha=1$ for $\mathbf{x} \in \Omega^e$ and $\alpha=1/2$ for $\mathbf{x} \in \partial\Omega$.

In the last integral, $\mathbf{F}(\mathbf{y})$ denotes the force exerted by $\partial\Omega$ on the suspending liquid, and must be determined from the boundary conditions. The kernels **J** and **K** correspond respectively to the single and double layer potentials, and are known functions of position given by:

$$J_{ij}(\mathbf{d}) = \frac{\delta_{ij}}{|\mathbf{d}|} + \frac{d_i d_j}{|\mathbf{d}|^3}, \qquad K_{ijk}(\mathbf{d}) = \frac{-3}{4\pi}\frac{d_i d_j d_k}{|\mathbf{d}|^5}, \qquad (4.10)$$

where $\mathbf{d} = \mathbf{x} - \mathbf{y}$. Further details on the boundary integral formulation may be found in the book by Pozrikidis [25].

4.2. Spherical capsule model.

In the case of an initially spherical capsule with radius A, undergoing small deformations (small shear rates or rigid membrane or high internal viscosity), it is possible to solve the problem by a perturbation method (Barthes-Biesel [26], Barthes-Biesel & Rallison [20], Barthes-Biesel & Sgaier [27]). The technique consists in expanding all quantities in terms of a small parameter and in obtaining successive approximations to the shape. When the capsule deformation is limited by the smallness of the capillary number C, to first order in C, the displacement of the membrane material points depend on two symmetric and traceless second-order tensors **J** and **K**, which are functions of time only:

$$\mathbf{x} = \mathbf{X} + C\{\mathbf{K}.\mathbf{X} + {}^T\mathbf{X}.(\mathbf{J}-\mathbf{K}).\mathbf{X}\ \mathbf{X}\} + O(C^2). \tag{4.11}$$

The in-plane deformations of the membrane elements are measured by **K**. The equation of the deformed profile of the particle depends only on **J** and is given by:

$$r = 1 + C\,{}^T\mathbf{x}.\mathbf{J}.\ \mathbf{x} + O(C^2). \tag{4.12}$$

Then, as shown by Barthes-Biesel & Sgaier, the solution to O(C) of equations (4.1) to (4.8) with any interfacial constitutive law (3.5) to (3.8) leads *in fine* to two differential equations defining the time evolution of **K** and **J**:

$$C\,\mathbf{K}^\circ = a_o\mathbf{E} + [b_o\mathbf{L} + b_1\mathbf{M}] - 2\beta\,[c_1\,\mathbf{J}^\circ + c_2\,\mathbf{K}^\circ]\,/c_o + O(C, \beta C)\,, \tag{4.13}$$

$$C\,\mathbf{J}^\circ = a_o\mathbf{E} + [b_o\mathbf{L} + (b_1 + b_2)\mathbf{M}] - 2\beta\,[c_3\,\mathbf{J}^\circ + c_4\,\mathbf{K}^\circ]\,/c_o + O(C, \beta C), \tag{4.14}$$

where the corotative Jaumann derivative is defined by:

$$\mathbf{K}^\circ = \partial\mathbf{K}/\partial t + \mathbf{K}.\mathbf{\Omega} - \mathbf{\Omega}.\mathbf{K}.$$

The parameter β represents the ratio between the membrane characteristic response time and the shear time. It is thus analogous to a Deborah number and is defined by:

$$\beta = \mu_S\, G / E_S.$$

The coefficients a_i , b_i and c_i are known functions of the viscosity ratio λ:

$$a_o = 5/(2\lambda + 3)\,; \qquad a_1 = 60(\lambda-1)/7(2\lambda+3)^2\,; \quad a_2 = 2(\lambda-1)/(2\lambda + 3);$$

$$b_o = 1/(2\lambda + 3); \qquad b_1 = 2(3\lambda + 2)/(19\lambda+16)(2\lambda+3); \quad b_2 = 2/(19\lambda+16)\,;$$

$$c_o = (19\lambda+16)(2\lambda+3); \quad c_1 = -2(7\lambda + 8)\,; \quad c_2 = 47\lambda+48;$$

$$c_3 = 2\,(\lambda+4)\,; \qquad c_4 = 15\lambda.$$

L and **M** are linear functions of **J** and **K**:

$$\mathbf{L} = 4\,(\alpha_2 + \alpha_3)\,\mathbf{J} - (6\alpha_2 + 10\alpha_3)\,\mathbf{K}, \tag{4.15}$$

$$\mathbf{M} = -4(2\alpha_2 + 2\alpha_3)\,\mathbf{J} + (12\alpha_2 + 16\alpha_3)\,\mathbf{K}, \tag{4.16}$$

in which the coefficients α_i depend on the constitutive equation of the membrane. Specifically, for a neo-Hookean membrane $\alpha_2 = 2\alpha_3 = 2/3$, while for the red blood cell membrane $\alpha_2 >> 1$ and $\alpha_3 = 1/4$.

This model presents some interesting general features. It accounts for the influence of the viscous deforming stresses due to any type of linear shear flow (terms involving **E** in (4.13) and (4.14)). It also accounts for the competing shape restoring effect due to elastic forces (terms involving **L** and **M**). Finally, the rotational 'tank-treading' motion of the membrane around a steady deformed shape is predicted. This effect is directly linked to the presence of the corotational derivatives in (4.14) and (4.13).

Once a flow field and a membrane constitutive behaviour are selected, it is very simple to solve equations (4.13) to (4.16), and to compute the deformed shape of the capsule. For example, in the case of a simple shear flow (2.2), the only non zero components of **J** are J_{11}, J_{22} and J_{12}. The profile is an ellipsoid with principal diameters L and B, oriented with respect to streamlines with an angle θ (Figure 4.1). For a linear visco-elastic law given by (3.6) and (3.9), Barthes-Biesel & Sgaier show that the deformation D in the shear plane and the angle θ are given by:

$$D_{12} = \frac{5C}{2\ (\beta^2 + 1)\ (\beta^2 + 1/9)}\,[\beta^2(\beta^2 + 5/9)^2 + \frac{1}{4}(\beta^2 + 5/9)^2]^{1/2}, \tag{4.17}$$

$$\tan 2\theta = \frac{\beta^2 + 5/9}{2\beta(\beta^2 + 7/9)}. \tag{4.18}$$

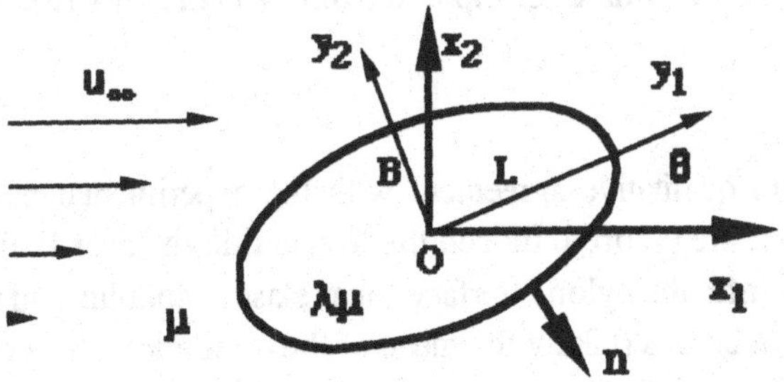

Figure 4.1 Schematics of the deformation of an initially spherical capsule in a simple shear flow

This analysis is valid provided that the membrane viscosity is large, i.e., the ratio $\mu A/\mu_s$ is small. In this limit, the spherical capsule model predicts that the deformation increases with shear rate up to a maximum value (Figure 4.2). The low shear and high shear asymptotic values, respectively D_0 and D_∞, are then found to be:

$$D_0 = \frac{25C}{4} = \frac{25\mu GA}{4E_s}, \qquad \text{and} \qquad D_\infty = \frac{5C}{2\beta} = \frac{5\mu A}{2\mu_s}.$$

Similarly, the capsule angle with respect to streamlines evolves from 45° to 0° as the shear rate increases.

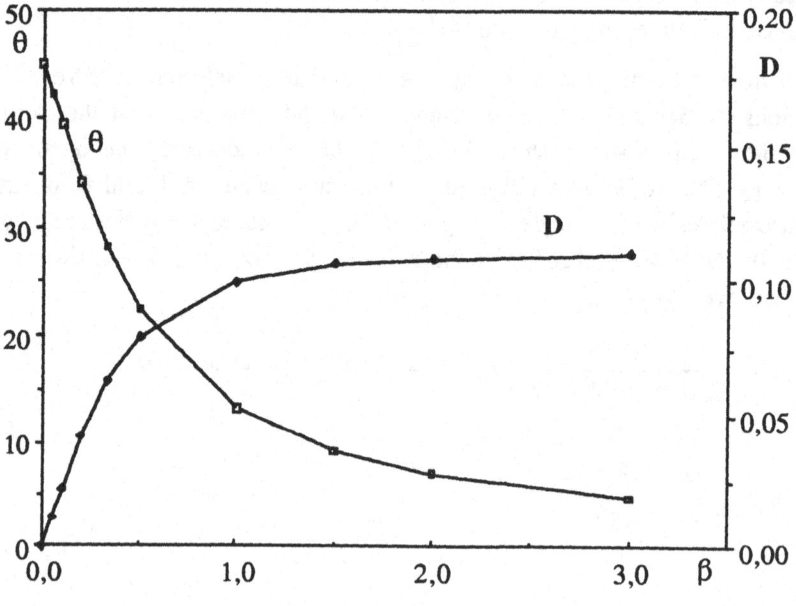

Figure 4.2 Deformation and orientation of an initially spherical capsule with a visco-elastic membrane. Computed from (4.17) and (4.18).

Those predictions are in qualitative agreement with the experimental results of Chang & Olbricht if the oscillations are ignored. Indeed the low and high shear limits have been used by Chang & Olbricht to infer the nylon interface shear elastic modulus and viscosity from the experimental deformation curves (taking the mean of the oscillation). The values obtained for E_S sometimes differ substantially from those computed from the squeezing experiments, however, the surface viscosity values are consistent with data on capsule relaxation after cessation of flow. The discrepancy observed for the E_S values might be attributed to

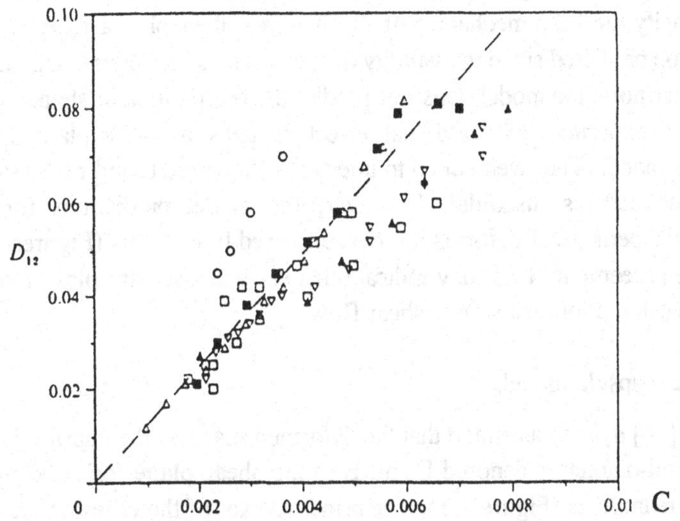

Figure 4.3 Deformation versus Capillary number for different capsules in extensional flow. The straight line is the prediction of Eq. (4.19). From Chang & Olbricht [7].

inhomogeneities in the capsule membrane that influence the squeezing measurements but are averaged out in simple shear flow.

In the case of the pure straining motion (2.3), the interfacial viscosity does not play any role since the membrane is stationary at steady state. Then the asymptotic low shear deformation is simply given by:

$$D = \frac{25C}{2} = \frac{25\mu GA}{2E_s} + O(C^2), \tag{4.19}$$

and no high shear limit is predicted. Equation (4.19) has been used by Akchiche and by Chang & Olbricht to analyze the low shear part of their deformation curves to infer the value of the membrane elastic modulus. The procedure is self consistent since data obtained for different capsules follow the same master curve (Figure 4.3).

It would thus appear that the spherical capsule model can be useful to interpret experimental observations of the motion and deformation of capsules subjected to flow, in terms of the interfacial mechanical properties. Other similar data are also pertinent, such as membrane rotation rate or relaxation behaviour under transient conditions. Finally it should be pointed out that the best experimental conditions correspond to cases where the viscosity ratio λ is of order 1 or less. Indeed, for large values of λ, the deformation process is essentially regulated by internal viscosity, and the mechanics of the interface then play a relatively minor role. Breakup cannot be predicted since the validity of the model is based on small deviations from sphericity. Furthermore, the model does not predict the oscillations in shape and orientation that are observed experimentally and that effect remains to be explained. Finally, the spherical capsule model is not well suited to interpret data on red blood cells, since the initial geometry of such cells is discoidal. However, the model predictions for visco-elastic capsules and the experimental deformation curve for red blood cells (Figures 2.6 and 3.2) are in qualitative agreement. This may indicate that surface viscosity plays a non negligible role in red blood cell motion in a simple shear flow.

4.3. Ellipsoidal capsule model.

Keller & Skalak [28] *apriori* assumed that the deformed shape of the capsule is an ellipsoid, with principal semi-diameters denoted L and B in the shear plane, oriented by an angle θ with respect to stream lines (Figure 4.1). The principal axes of the ellipsoid are denoted Oy_1 and Oy_2, with corresponding unit vectors $\mathbf{e}'_1$ and $\mathbf{e}'_2$. A simple membrane velocity is then assumed to reproduce the motion of a tank-treading red blood cell:

$$\mathbf{u}_S = R_F \left(L/B \, y_2 \, \mathbf{e}'_1 - B/L \, y_1 \, \mathbf{e}'_2 \right), \tag{4.20}$$

where R_F is the rotational frequency of the membrane. This velocity field respects the overall membrane inextensibility, but is not locally area preserving. The Stokes equations for the external flow are solved subject to boundary conditions (4.6) and (4.7). The velocity distribution (4.20) is linearly extended in the interior of the cell, and the corresponding rate of energy dissipation is computed. The problem is closed by assuming that the membrane does not dissipate energy and that the external liquid energy is all dissipated by the internal motion. The model predicts two types of behaviour for the cell: an unsteady flipping motion or a steady orientation with a tank-treading membrane. The transition from the first behaviour to the second one occurs when the viscosity ratio λ is decreased and/or when the cell elongation L/B is increased. This is in good qualitative agreement with rheoscope observations of a red blood cell. However, it should be noted that in this model, the cell geometry and the viscosity ratio both are considered as two independent parameters in spite of the fact that they are linked by the mechanics of the system. This is a consequence of the fact that the membrane equations are not solved exactly.

This tank-treading ellipsoid model is used by Sutera and his coworkers, in a series of papers ([29],[30],[31]), to analyze the motion of a red blood cell in the shear flow created in a rheoscope. After the internal and external velocity fields are determined as indicated above, it is simple to compute the stress distribution in the fluids from Eq. (4.4) and (4.5). The load $\mathbf{q}$ on the membrane is given by Eq. (4.8), and the elastic tensions can be obtained directly from Eq. (3.4) since the system is statically determinate. In the case of a visco-elastic membrane (Eq. (3.8) and (3.9)), it is possible to compute the shear elastic modulus G_s and surface viscosity μ_s. The ellipsoidal tank-treading capsule model thus allows to deduce the membrane properties of a red blood cell from rheoscope observations. The model predicts a slight shear thinning behaviour for G_s and μ_s which has yet to be validated by other measurements.

4.4. Large deformation of a capsule in elongational flow.

The present trend is to develop numerical models, that can allow for large deformations and for capsules with an initial arbitrary geometry. The integral formulation of the Stokes equations (Eq. 4.9) is convenient, and the problem is usually solved by means of a numerical collocation method. However, such models are very complicated to design and demand large computing times. Consequently, most of the presently available models deal with axisymmetric situations where the flow and the capsule are axisymmetric. They thus only provide some qualitative insight on the mechanics of capsules suspended in actual straining motions.

The case of spheroidal capsules, with a Mooney-Rivlin membrane (Eq. 3.6), suspended in a pure straining flow has been considered by Li et al. [32]. It is found that the deformation increases with capillary number, until a critical value of C is reached, past which there exists no steady state solution to the equations of motion (Figure 4.4). For values of C larger than the critical level, the capsule elongates without bound until obviously a failure criterion for the membrane is eventually reached. Burst of the particle is thus predicted according to what is found experimentally for a polylysine capsule. The critical value of C depends on the parameters Ψ and Ψ', and also on the initial shape of the capsule. In the case of a spherical capsule with a neo-Hookean membrane, this critical value is of order 0.085. Results have also been obtained for initially spherical capsules (Barthes-Biesel [11]) with a strain-hardening membrane of the Hart-Smith type such that:

$$T_{11} = \frac{E_s}{3\lambda_1\lambda_2}\left[\lambda_1^2 - \frac{1}{(\lambda_1\lambda_2)^2}\right], \qquad (4.21)$$

with $E_s = E_{s0} \exp\{ K_s [\lambda_1^2 + \lambda_2^2 + (\lambda_1\lambda_2)^{-2} - 3]\}$.

The Capillary number is based on E_{s0}. The non-dimensional parameter K_s measures the amount of strain hardening. The stiffening effect under increasing levels of strain limits the deformation of the capsule and prevents it from bursting through the continuous elongation process (Figure 4.4). This agrees with what is observed experimentally for a nylon capsule.

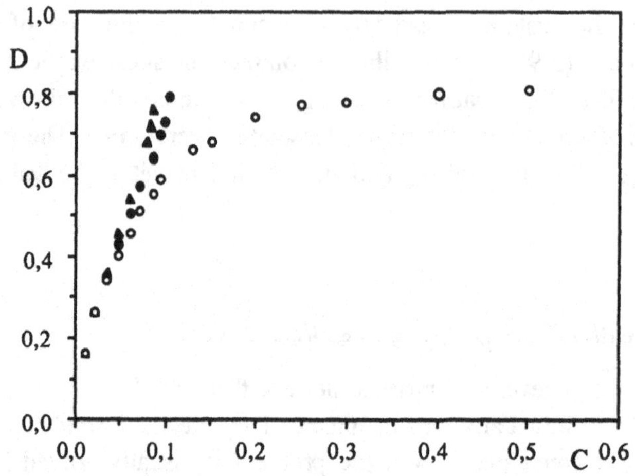

Figure 4.4 Deformation versus Capillary number of an initially spherical capsule with a neo-Hookean membrane (▲), with a strain hardening membrane (●) K_s =0.05, (o) K_s= 0.25.

A similar numerical model has also been devised by Pozrikidis [33] for capsules with an area incompressible membrane described by Eq. (3.8). In that case, the area incompressibility limits the particle deformation and prevents break-up.

4.5. Conclusion.

The motion of capsules in unbounded shear flows represents a difficult problem where solid and fluid mechanics are coupled. The analytical models developed for spherical moderately deformed capsules work reasonably well to interpret experimental data in terms of the particle intrinsic properties. Recently Pozrikidis [34] and Zhou & Pozrikidis [35] have presented the first 3D numerical study of a capsule in simple shear flow, in the case where the internal and external viscosities are equal (λ=1). This model is useful to determinate the range of validity of the perturbation solution (Eq. 4.19), which is of order C=0.05. It is also very useful to study the role of the initial geometry of the capsule as it can allow for non spherical shapes. Furthermore, the model predicts burst of capsules with a Mooney-Rivlin type membrane.

5. MOTION OF A CAPSULE IN A PORE.

The objective is to try and model situations where a capsule has to flow into a tube with a diameter r_t smaller or of the same order as the capsule typical dimension. The case where the capsule is much smaller than the tube diameter is less interesting since the hydrodynamic perturbation it creates is small. However, the situation where a small initially spherical capsule is suspended in an unbounded parabolic flow ($A/r_t >> 1$) has been considered by Barthes-Biesel & Helmy [36]. Then a perturbation method similar to the one presented in section 4.2, may be devised. It is found that the capsule migrates towards the tube axis. This phenomenon will contribute to the creation of a cell free layer near the tube wall during suspension flow.

5.1. Problem formulation.

The problem that must be addressed is the one described by equations (4.1) to (3.11), with additional conditions that arise from the presence of the channel walls and from input and output conditions. The external flow domain is delimited by solid boundaries B, corresponding to the pore, and by two entrance and exit sections, respectively S_1 and S_2, where the velocity profiles $\mathbf{u}_1(\mathbf{x})$, $\mathbf{u}_2(\mathbf{x})$ and pressures p_1, p_2 are specified (Figure 5.1). To the set of equations (4.6), (4.7), (4.8) additional boundary conditions must be added:

$$\mathbf{u}(\mathbf{x}) = \mathbf{0}, \qquad \mathbf{x} \text{ on } B, \tag{5.1}$$

$$\mathbf{u}(\mathbf{x}) = \mathbf{u}_i(\mathbf{x}), \text{ and } p(\mathbf{x}) = p_i, \qquad \mathbf{x} \text{ on } S_i, \ i = 1,2. \tag{5.2}$$

The sections S_1 and S_2 are taken far enough from the particle, for the flow perturbation to be negligible. Depending on the flow configuration, either the total pressure drop ΔP or the total flow rate Q across the pore is kept constant:

- constant pressure drop:

$$p_1 - p_2 = \Delta P_0, \qquad Q(t) = Q_0 - \Delta P^+(t)/R_H, \tag{5.3}$$

- constant flow rate:

$$p_1 - p_2 = \Delta P(t) = \Delta P_0 + \Delta P^+(t), \qquad Q(t) = Q_0, \tag{5.4}$$

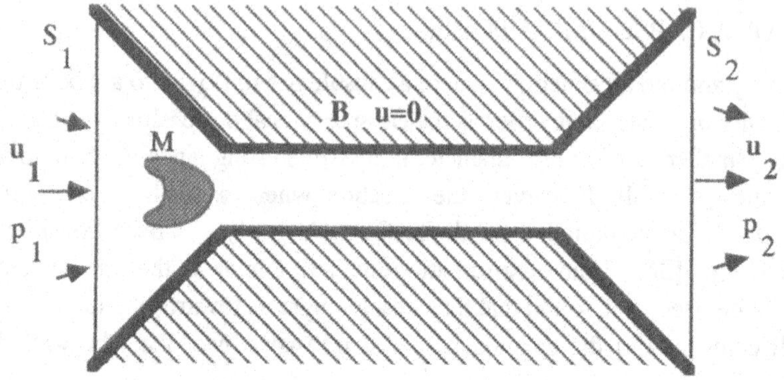

Figure 5.1 Schematics of the flow of a capsule through a pore.

where $\Delta P^{+}(t)$, denotes the additional pressure drop due to the presence of the cell and where R_H is the pore hydraulic resistance that depends only on pore geometry. The Capillary number is then redefined as :

$$C = \frac{\mu\, Q_0}{\pi\, r_t^2\, E_s} \quad . \tag{5.5}$$

5.2. Tightly fitting capsules.

Apart from the case of for small spherical capsules in large pores, this problem can also be resolved analytically for tightly fitting particles in cylindrical tubes (Halpern & Secomb [37]). Then, the lubrication approximation is used to simplify the equations of motion of the fluids and to determine the flow in the thin liquid film that surrounds the capsule. Similarly, the case of channels with slowly varying cross section and closely fitting cells can also be treated [38], provided that the hypotheses inherent to lubrication are fulfilled. Flaccid capsules with an area incompressible membrane (Eq. 3.8) have been specifically considered as a model of blood cell in microcirculation [37], [38]. It is then found that the cell takes a slug shape with a long cylindrical central part. The front is always convex and extended, whereas the back is concave or convex depending on the Capillary number and on the size ratio between the tube and the capsule. In other intermediate situations, a numerical solution of the problem must be sought. As a matter of fact, the closure of the asymptotic problem for tightly fitting particles, is also numerical.

5.3. Numerical models of the flow of a capsule through a pore.

Up to now, only fully axisymmetric situations have been considered where the channel and the capsule share the same revolution axis. The integral form (Eq. 4.9) of the Stokes equations is used, and the motion of a capsule as it flows into and through the pore is followed by means of a forward time stepping method. The case of a short pore has been considered by Leyrat-Maurin [39] and by Leyrat-Maurin & Barthes-Biesel [40]. The geometry of the flow channel B is that of axisymmetric hyperboloid, with throat radius r_t, and such that the angle of the hyperbola asymptote with Ox is 45°. In this situation, entrance and exit effects prevail and the flow is essentially transient. Similarly, the case of long pores with hyperbolic entrance and exit sections (with the same geometry as in [40]) has been studied by Quéguiner [41] and by Quéguiner and Barthes-Biesel [42]. In this situation after an initial transient due to entrance effects, the capsule reaches a steady shape. Results have been obtained for the case of initially spherical capsules surrounded by a neo-Hookean type membrane (Eq. 3.6), and filled with an internal liquid that has the same viscosity as the suspending medium ($\lambda = 1$). An important new parameter then appears, which corresponds to the size ratio between the particle and the tube:

$$R = A/r_t.$$

The two models allow the study of the transient motion and deformation of a cell through the pore. The time evolution of local quantities is calculated, such as particle geometry, center of mass velocity, and elastic tensions in the membrane. The cylindrical tube model allows to predict steady state values that will occur in a long enough pore.

The successive profiles a small capsule (R=0.8 and C=0.04) in a cylindrical pore are shown on Figure 5.2: the capsule takes an oblong shape during the entrance process. However, this shape is not in equilibrium with Poiseuille flow, and the steady profile that is finally obtained is "parachute" like, and looks very much like the red blood cell shape in the human microcirculation. During exit the front of the particle slows down and this leads to an increase of its radial dimension. A large capsule such that R=1.2, C=0.005, shows a tendency to form slugs (Figure 5.3). The rear of the capsule is convex for small C, and becomes concave as C increases. The liquid film around the particle is thin. The flow in this region could also be treated by means of the lubrication approximation, with proper matching with the two extremities. The same phenomenon of radial expansion is also noted during exit. It may be important enough for the capsule membrane to almost touch the pore wall. This occurrence of exit plugging was first predicted by Leyrat-Maurin & Barthes-Biesel in the case of an hyperbolic channel. It is known to occur during filtration experiments on blood, where clusters of red blood cells attached to the downstream side of a filter have been reported [43].

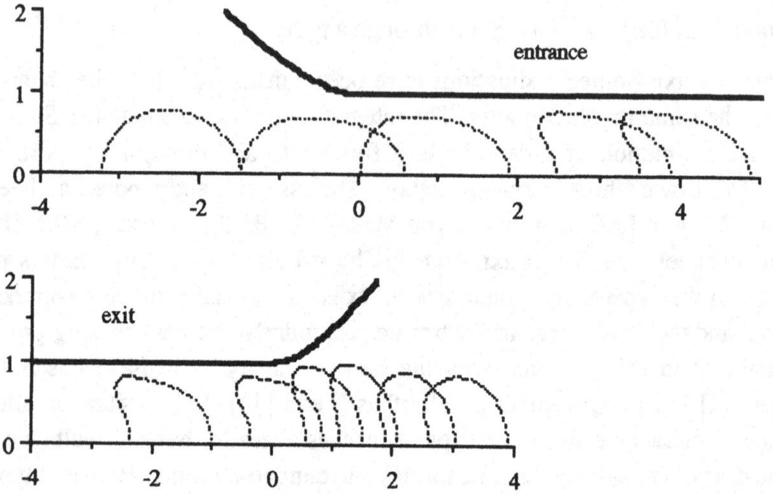

Figure 5.2. Successive profiles of a small capsule during motion through a tube (R=0.8, C=0.03). From [41].

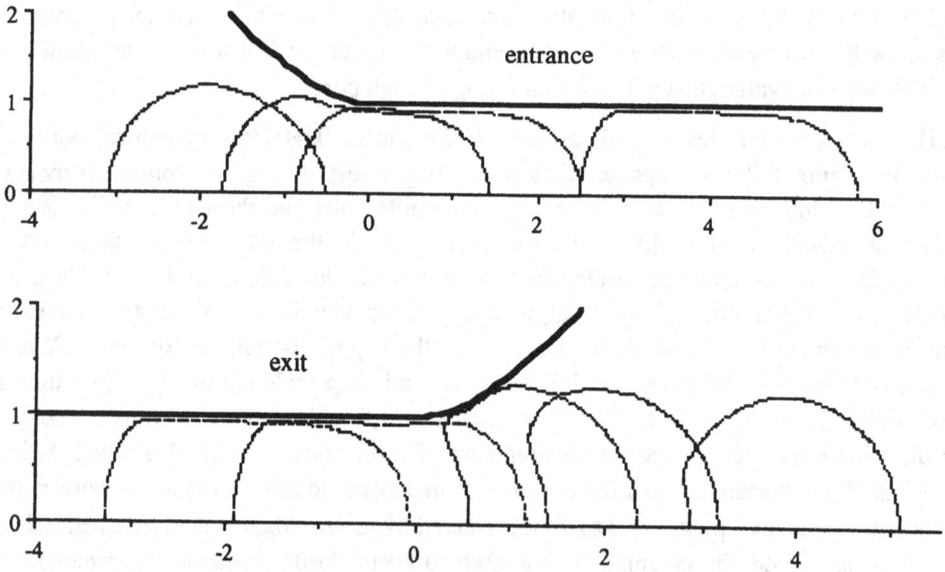

Figure 5.3. Successive profiles of a small capsule during motion through a tube (R=1.2, C=0.005). From [41].

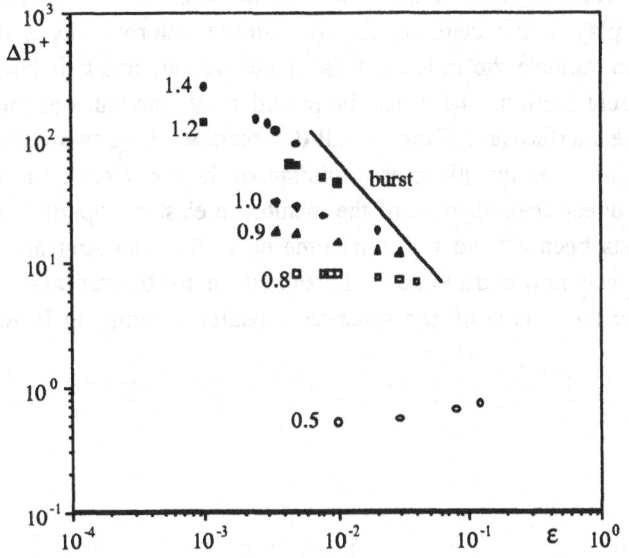

Figure 5.4 Steady additional pressure drop for spherical neo-Hookean capsules. The numbers refer to the values of R. From [41].

The effect of capillary number and capsule size on the steady additional pressure drop can be predicted (Figure 5.4). When C is increased past a critical value that depends on R, no steady state solution can be obtained and the capsule seems to deform continuously. It appears that this model predicts burst of a capsule for large enough flow strengths. Burst then occurs because elastic forces can no longer balance the viscous forces imposed by the flow. For a given capsule size, the steady state value of the additional pressure drop ΔP^+ decreases when C, i.e. deformability, increases. For small capsules, this dependency is mild, however, it is much more pronounced for large capsules. Altogether, the model predicts that the additional pressure drop depends strongly on size, but mildly on membrane properties as measured by C. This models thus corroborates the experimental evidence regarding the influence of membrane properties on filtration parameters (section 2).

6. CONCLUSION.

There is still much work to be done to understand the respective roles that the different intrinsic parameters play in the determination of capsule deformability and resistance to stress. To be able to validate the models, it is of course important to have independent measures of the capsule motion. These may be provided by simultaneous measurements of the rotation rate of the interface and of the overall deformation along two different directions (presently only two of these quantities are simultaneously measured). Furthermore, it is necessary to have a direct determination of the membrane elastic properties. Except for the micropipet, which has been around for some time now, flat film stretching or squeezing experiments are relatively new and should be developed. On the theoretical side, a general 3-dimensional model of the motion of non spherical capsules undergoing large deformations should be developed.

REFERENCES.

1. Fung Y.C. (1981) Biomechanics, chapt. 4. New-York, Springer-Verlag.
2. Skalak R., Özkaya N., Skalak T.C. (1989) Biofluid mechanics. Ann. Rev. Fluid Mech., **21**, 167-204.
3. Hochmuth R.M., Berk D.A. (1984) Analytical solutions for shear deformation and flow of red cell membrane. J. Biomech. Eng., **106**, 2-9.
4. Hochmuth R.M., Waugh R.E. (1987) Erythrocyte membrane elasticity and viscosity. Ann. Rev. Physiol., **49**, 209-219.
5. Lardner T.J., Pujara P. (1980) Compression of spherical shells. Mech. Today, **5**, 161-176.
6. Hiramoto Y. (1970) Rheological properties of sea-urchin eggs. Biorheology, **6**, 201-234.
7. Chang K.S., Olbricht W.L. (1993) Experimental studies of the deformation of a synthetic capsule in extensional flow. J. Fluid Mech., **250**, 587-608.
8. Chang K.S., Olbricht W.L. (1993) Experimental studies of the deformation and breakup of a synthetic capsule in steady and unsteady simple shear flow. J. Fluid Mech., **250**, 609-633.

9. Burger A., Rehage H. (1992) From two-dimensional model networks to microcapsules, Ang. Makromol. Chem., **202-203**, 31-44.

10. Akchiche M. (1987) Thèse de Doctorat. Université de Compiègne.

11. Barthes-Biesel D. (1991) Role of interface properties on the motion and deformation of capsules in shear flow. Physica A., **172**, 103-124.

12. Bentley B.J., Leal L.G. (1986) An experimental investigation of drop deformation and breakup in steady, two-dimensional linear flows. J. Fluid Mech., **167**, 241-283.

13. Schmid-Schonbein H., Wells R.E. (1969) Fluid to drop like transition of erythrocytes under shear. Science, **165**, 288-291.

14. Pfafferott C., Wenby R., Meiselman H.J. (1982) Morphologic and internal viscosity aspects of red blood cell rheologic behaviour. Blood Cells, **8**, 68-78.

15. Nash G.B. (1990) Filterability of blood cells: methods and clinical applications. Biorheology, **27**, 873-882.

16. Fischer T.C., Wenby R.B., Meiselman H.J. (1992) Pulse shape analysis of RBC micropore flow via new software for the cell transit analyser (CTA). Biorheology, **29**, 185-201.

17. Kieswetter H., Dauer U., Teitel P., Schmid-Schonbein H., Trapp R. (1982) The single erythrocyte rigidometer (SER) as a reference for RBC deformability. Biorheology, **19**, 737-753.

18. Schmid-Schonbein H., Gaehtgens P. (1981) What is red cell deformability? Scand. J. Clin. Lab. Invest., **41**, Suppl.156, 13-26.

19. Drochon A., Barthes-Biesel D., Bucherer C., Lacombe C., Lelievre J.C. (1993) Viscous filtration of red blood cell suspensions.Biorheology, **30**, 1-7.

20. Barthes-Biesel D., Rallison J.M. (1981) The time dependent deformation of a capsule freely suspended in a linear shear flow. J. Fluid Mech., **113**, 251-267.

21. Skalak R., Tozeren A., Zarda R.P., Chien S. (1973) Strain energy function of red blood cell membranes. Biophys. J., **13**, 245-264.

22. Evans E.A. (1973) A new material concept for the red cell membrane. Biophys. J., **13**, 926-940.

23. Brunn P.O. (1983) The deformation of a viscous particle surrounded by an elastic shell in a general time-dependent linear flow field. J. Fluid Mech. **126**, 533-544.

24. Rallison J.M. & Acrivos A. A numerical study of the deformation and burst of a viscous drop in an extensional flow.J. Fluid Mech., 1978, 89, 191-200.

25. Pozrikidis C. Boundary integral and singularity methods for linearized viscous flow. Cambridge University Press, 1992.

26. Barthes-Biesel D.(1980) Motion of a microcapsule in shear flow J. Fluid Mech.,**100**, 831-853.

27. Barthes-Biesel D., Sgaier H. (1985) Role of membrane viscosity in the orientation and deformation of a capsule suspended in shear flow. J. Fluid Mech., **160**, 119-135.

28. Keller S.R., Skalak R. (1982) Motion of a tank-treading ellipsoidal particle in a shear flow. J. Fluid Mech., **120**, 27-47.

29. Tran-Son-Tay R., Sutera S.P., Rao P.R. (1984) Determination of RBC membrane viscosity from rheoscopic observations of tank-treading motion. Biophys. J., **46**, 65-72.

30. Tran-Son-Tay R., Sutera S.P., Zahalak G.I., Rao P.R. (1987) Membrane stress and internal pressure in a RBC freely suspended in a shear flow. Biophys. J. **51**, 915-924.

31. Sutera S.P., Pierre P.R., Zahalak G.I. (1989) Deduction of intrinsic mechanical properties of the erythrocyte membrane from observations of tank-treading in the rheoscope. Biorheology **26**, 177-197.

32. Li X.Z., Barthes-Biesel D. & Helmy A. (1988), Large deformations and burst of a capsule freely suspended in an elongational flow. J. Fluid Mech. 187, 179-196.

33. Pozrikidis C. (1990) The axisymmetric deformation of a red blood cell in uniaxial straining Stokes flow. J. Fluid Mech., **216**, 231-254.

34. Pozrikidis C. 1995 Finite deformation of liquid capsules enclosed by elastic membranes in simple shear flow. J. Fluid Mech., **297**, 123-152.

35. Zhou H. & Pozrikidis C. (1995) Deformation of liquid capsules with incompressible interfaces in simple shear flow. J. Fluid Mech. **283**, 175-200.

36. Helmy A. and Barthès-Biesel D. (1982) Migration of a spherical capsule freely suspended in an unbounded parabolic flow. J. de Mécanique Théorique et Appliquée. **1**,859-880.

37. Halpern D. & Secomb T.W. (1989) The squeezing of red blood cells through capillaries with near minimal diameters. J. Fluid Mech. , **203**, 381-400.

38. Secomb T.W., Skalak R., Özkaya N. & Gross J.F. (1986) Flow of axisymmetric red blood cells in narrow capillaries. J. Fluid Mech. , **163**, 405-423.

39. Leyrat-Maurin, A. (1993) Thèse de Doctorat . Université de Compiègne.

40. Leyrat-Maurin, A., Barthes-Biesel, D. Motion of a spherical capsule through a hyperbolic constriction. J. Fluid Mech., 1994, **279**, 135-163.

41. Quéguiner C. (1995) Thèse de Doctorat . Université de Compiègne.

42. Quéguiner C., Barthes-Biesel D. (1996) Flow of capsules through cylindrical channels. Submitted to J. Fluid Mech.

43. Chien S., Schmid-Schönbein G.W., Sung K.L.P., Schmalzer E.A. and Skalak R.(1984) Viscoelastic properties of leukocytes in White Cell Mechanics: basic science and clinical aspects, Alan R. Liss Inc 1984, 19-51.

THE EFFECT OF SURFACTANTS ON THE MOTION OF BUBBLES AND DROPS

Ch. Maldarelli and Wei Huang
City College of New York, New York, NY, USA

ABSTRACT

When bubbles or drops move and deform through a continuous liquid phase as in suspension flows, and a surface active species (surfactant) is present in the suspension, the surfactant absorbs onto the fluid particle interfaces. Adsorption lowers the surface tension in proportion to the surface concentration. Surfactant adsorbed onto the surfaces of moving fluid particles is partitioned by the surface flow and the interfacial deformation,creating surface tension gradients. The gradients exert a Marangoni traction on the continuous phase which affects the suspension flow. In this chapter we present a basis for the understanding and fluid mechanical modeling of the influence of Marangoni stresses. Attention is focused on the air/water inerface and the motion af bubbles.

The chapter first describes the phase behavior of surfactant monolayers, the experimental methods used to identify phase polymorphism, and the principal phases for bulk soluble surfactants: gaseous, liquid expanded, liquid condensed and solid. Next, we describe the modeling of the state equations for the dependence of the tension on the surface concentration, and the rate equations for the exchange of surfactant between the surface and the adjoining sublayer. The measurement of the equation of state parameters and kinetic rate constants is described next, and we examine the surfactant systems which have been studied. The chapter concludes with a study of the buoyancy driven, inertialess motion of a bubble in a continuous liquid phase, as a model problem for the illustration of the inclusion of Marangoni stresses in a fluid particle flow. We compute the terminal velocity as a function of the surfactant transport properties.

1. INTRODUCTION

The flow of suspensions consisting of bubbles and drops - as, for example, the motions of foams and emulsions - is affected by the presence of surfactant molecules dissolved in the liquid phases of the multiphase system. The aim of this chapter is to provide a framework for the understanding and fluid mechanical modeling of the influence of surfactants on fluid particle motions. We first provide in this Introduction some background information on the structure and physical chemistry of surfactants, and a discussion of the basic mechanisms by which surfactants affect the motion of fluid interfaces, and the mathematical formalisms for accounting for these mechanisms.

1.1 Surfactant Structure and Physical Chemistry

A surfactant is an amphiphillic molecule consisting of a nonpolar or hydrophobic group and a polar or hydrophilic group which are spatially disjoint from one another in the molecule. The nonpolar group (denoted by **R**) is usually a string of hydrocarbon methylene groups ($-(CH_2)-$) which are arranged in one (or more) chains each of which may be branched or unbranched. The polar group can be uncharged as a hydroxyl (**R**-OH), or polyoxyethylene ((**R**-$(-OCH_2CH_2-)_n$-OH) group, anionic as a carboxylate (**R**-COO^-), sulphate (**R**-SO_3^-) or sulphonate group (**R**-OSO_3^-), cationic as a quartenary ammonium group (**R**-$NR'_3{}^+$), or amphoteric (or zwitterionic) in which the polar group contains a negative and positive group as sulfbetaines (**R**-$N^+(CH_3)_2CH_2CH_2SO_3^-$).

Surfactants dissolved in fluid systems consisting of two immiscible phases of significantly different polarity have a strong tendency to adsorb at the interfaces between the phases. Surfactants relocate at fluid interfaces because at these interfaces they can configure themselves so that the polar moiety remains immersed in the aqueous phase (thereby enabling the strong interaction between the polar group and water) and the nonpolar moiety is positioned in the air or the hydrophobic phase. The adsorption lowers the free energy of the system, and reduces the surface or interfacial tension γ of the fluid interface in proportion to the amount adsorbed (or surface concentration, Γ). The *equation of state* describes the dependence of the surface tension on the surface concentration ($\gamma(\Gamma)$). The *adsorption isotherm* describes the dependence of the surface concentration on the bulk concentration, C ($\Gamma(C)$). Both the equation of state and adsorption isotherm strictly refer to equilibrium states. As the bulk concentration of surfactant is increased, the adsorption at the surface becomes saturated, and additional surfactant dissolved in the liquid phases can form a variety of bulk aggregate structures or micelles. Micelles are arranged so as to shield the part of the surfactant incompatible with the liquid phase; thus for surfactants adsorbed in water, one micellar structure is a spherical aggregate in which the chains cohere together in a core surrounded by a corona of the polar groups. A comprehensive introduction to the chemistry of surfactant molecules is given in the text by Rosen [1]

The exchange of surfactant between a surface and a liquid sublayer immediately adjacent to the surface is a kinetic process consisting of adsorption and desorption steps which are usually described by rate equations. This interfacial transport is distinguished from the bulk diffusion of surfactant. We illustrate this difference in Fig. 1.1 below which details the in-series processes of kinetic adsorption and bulk diffusion for adsorption of surfactant onto an initially clean interface.

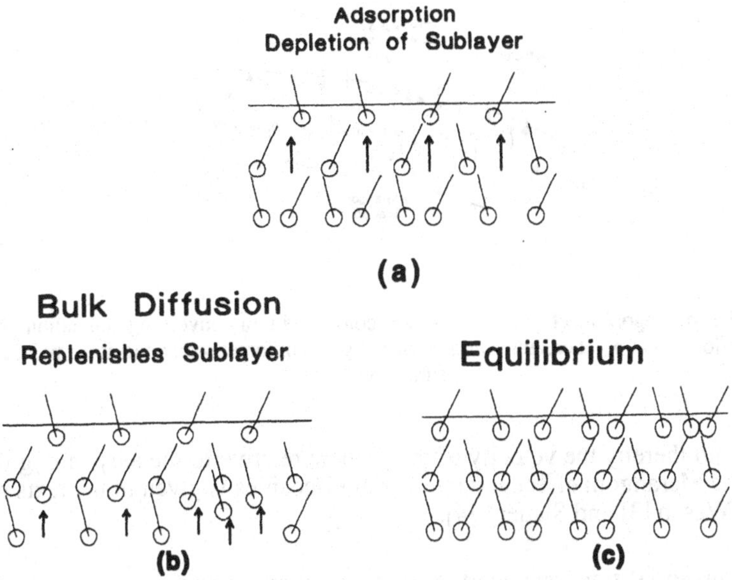

Fig. 1.1, Surfactant transport to an initially clean interface. First, kinetic adsorption reduces the sublayer concentration (a). Second, sublayer depletion drives a diffusive flux which supplies surfactant to the sublayer (b). Third, diffusive flux replenishes the sublayer until equilibrium is achieved (c).

Kinetic and diffusive transport of surfactant is reviewed in the survey article of Chang and Franses [2].

1.2 The Marangoni Mechanism

When surfactants adsorb onto the surfaces of moving bubbles or drops separating phases of different polarity (as an air bubble or oil drop in water), they affect the fluid particle mechanics in two fundamental ways. First adsorption lowers the interfacial tension of the particle interface, and thereby increases the deformability of the particle. Second, adsorption retards the interfacial mobility of the surface, i.e. the ability of the surface to move relative to the motion of the particle's center of mass. Two mechanisms account for this reduction in interfacial mobility. The first is the Marangoni effect: When bubbles or drops move, interfacial flow sweeps fluid to converging stagnation points on the surface. For example, for a bubble moving in a continuous phase, surfactant is convected from the leading pole to the trailing pole. Surfactant adsorbed at the interface is swept to, and accumulates at, the regions on the surface where the interfacial flow converges. Accumulation lowers the tension in these converging regions relative to the tension of the surroundings interface, and the surrounding interface therefore tugs at the converging zone and retards the surface flow (Fig.1.2). The second mechanism is a direct surface rheological effect: Interactions among surfactant molecules in an adsorbed layer on an interface creates resistance to shear and dilatational flow along the interface. We note however that for bulk soluble surfactants of low molecular weight and reasonably high bulk solubility, surface viscosities are not usually as important as Marangoni effects because the surface monolayers are not strongly cohered. The reduction in interfacial mobility increases the drag exerted by the continuous phase on

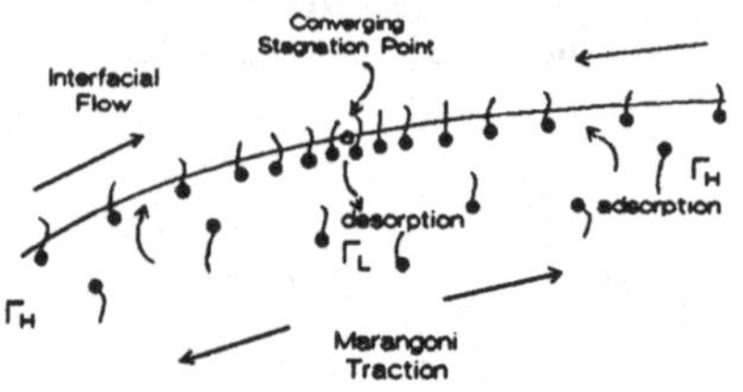

Fig. 1.2, The Marangoni mechanism. Surfactant convected to a converging stagnation point locally reduces the tension at that point creating surface tension gradient forces which tug at the point and reduce the surface velocity.

the particle, and thereby the velocity of the particle relative to the suspending phase. Introductions to the Marangoni effect and interfacial viscosities is given in the texts by Edwards, Brenner and Wasan [3] and Slattery [4].

The tangential tractions exerted by the gradient in surface tension (γ) balance the action of the viscous stresses exerted on the interface from the adjoining phases:

$$[F_t] = \nabla_s \gamma = \frac{\partial \gamma}{\partial \Gamma} \nabla_s \Gamma \tag{1.1}$$

where ∇_s is the surface gradient operator, $\mathbf{F}_t$ is the tangential viscous traction vector exerted by the bounding phases on the interface, and the bracket denotes the difference evaluated across the interface. The second equality in (1.1) follows from the surfactant equation of state, which expresses the dependence of the tension (γ) on the surface concentration (Γ). The surface density is computed from the equation for the conservation of mass at the interface:

$$\frac{\partial \Gamma}{\partial t} = -\nabla_s \bullet (\Gamma V_t) + Q + 2H\Gamma V \bullet n + X \bullet \nabla_s \Gamma \tag{1.2}$$

In the above, the time derivative is evaluated with the surface coordinates held fixed, $\mathbf{V}_t$ is the tangential velocity vector at the surface, H is the mean curvature, **X** is the velocity of a point on the surface which retains the same surface coordinate, **n** is the normal vector to the surface, and **V** is the surface velocity vector. Finally, Q is the kinetic rate of exchange of surfactant between the surface and the sublayers adjoining the surface. This rate is equated to the net rate of diffusion of surfactant normal to (and evaluated at the surface:

$$Q = n \bullet [D \nabla C] \tag{1.3}$$

where we have used Fick's law to express the diffusional flux, the bracket again indicates the difference across the interface and D denotes the diffusion coefficient. The above equations are discussed particularly well in the reference monograph of Slattery [4] .

1.3 Scope

The above discussion in **1.1** and **1.2** indicates that to properly understand and model the effect of surfactant on the fluid particle motion, the equation of state $\gamma(\Gamma)$, and the kinetic exchange rate, Q must be specified (and the diffusion coeffiecient D measured). Only then can the proper hydrodynamic modeling be undertaken. In order to develop equations of state and kinetic models one must first understand the kinds of surface phases in which a monolayer can exist. We therefore begin our study in this chapter with Sec. **2** which examines the different monolayer phases. In this and subsequent sections, we will restrict our attention to the air/water interface since this is the most thoroughly studied. In the following section (**3**) we review theoretical models for the equation of state and the kinetic rate relations. We explain how the model constants in the state and kinetic equations are obtained experimentally in Sec. **4**. Many hydrodynamic problems connected with the motion of bubbles and drops with adsorbed surfactant monolayers have been examined in the literature. Among them, with special application to the flow of suspensions, are the motion of single drops or bubbles in a continuous, otherwise quiescent liquid phase, the coalescence of bubbles or drops, the extension of drops in straining flows and the motion of bubbles and drops in cylindrical tubes. In this chapter, to illustrate the hydrodynamic modeling of bubble and drop motions with surfactant laden interfaces, we choose the simplest example of bubble motion in a quiescent liquid and discuss this in Sec. **5**. Studies of bubble coalescence in the presence of surfactant are reviewed in the monographs by Ivanov [5], Edwards [3] and Slattery [4] and [6-13]. The effect of surfactant on straining flows is given in [14-16], and the motion of fluid particles in tubes is treated in [17-21].

2. MONOLAYER PHASE BEHAVIOR AT THE AIR / WATER INTERFACE

When spread at the air/water interface, monolayers of water ***insoluble,*** single chain amphiphiles with simple polar groups can exist in different phases or states, depending on the surface density (or the surface pressure), temperature, and surfactant structure. A host of techniques, including surface pressure measurements [22], fluorescence[23] and Brewster angle microscopy [24, 25] studies, and x-ray [26], electron diffraction, [27] and neutron reflectivity investigations, have distinguished these monolayer states in terms of molecular ordering including the positional (translational) order of the head groups and the orientational (tilt) order of the chains. These methods have also located the boundaries between the states in surface pressure (or area per molecule) / temperature space, and have established whether the transitions at the boundaries are first order, indicating phase coexistence, or higher order [28-30]. Each phase represents the balance, for a particular temperature and surface density or pressure, between the cohesive van der Waals interactions of the chains which favor cohered states of chains stacked lengthwise, and the repulsive interactions of the head group (dipolar, electostatic and hydration interactions) which favor more expanded, less ordered states. The phase behavior of monolayers of water ***soluble*** surfactants, assembled by adsorption from solution has not received as much attention. We first briefly review the current state of understanding of insoluble monolayer phases, focusing on chain lengths comparable with those found for water soluble surfactants (less than 20 carbon atoms), and then we discuss the less understood surface states of soluble amphiphiles.

Polymorphism in insoluble monolayers is evident in isothermal surface pressure/area per molecule (Π-A) measurements, in which phase changes are demarcated either by horizontal lines indicating coexistence and a first order transition, or kinks indicating transitions of higher order between phases of different compressibility χ ($\chi = -\, d\ln A/d\Pi$). The most carefully studied are the alkanoic or fatty acids, and of these pentadecanoic acid (PDA) represents the medium chain length (C_{14}-C_{18}) archetype, showing up to four states and having two first order phase transitions.

At very low surface densities, the PDA monolayer exists in a dilute gaseous phase (G), having low surface pressures. As the area per molecule is reduced from 1500 to 42 Å^2/molecule, an extended surface pressure plateau of 0.132 mN/m is found at 20 °C, indicating emergence and coexistence of a disordered liquid expanded phase (denoted by LE or L_1) with gaseous regions, through a first-order phase transition [31]. At the high density end of the transition, the entire interface is in the LE state, and further reduction in the area raises the surface pressure, with a compressibility of ~ 0.08 m/mN. Between 35 and 24 Å^2/molecule, the surface pressure is again constant, at 2.5 mN/m, indicative of coexistence between disordcred (LE) and condensed tilted liquid states (referred to as LC or L_2). Below 24 Å^2/molecule, the pure LC interface can be compressed slightly (χ ~ 0.01 m/mN),[32] with a steep rise in Π through the equilibrium spreading pressure of ~18 mN/m.[33] Overcompression of the monolayer shows a kink in the surface pressure at 21 mN/m (20 Å^2/molecule) reflecting a transition to a (metastable) vertical solid state (LS), [34] with $\chi \sim 10^{-3}$ m/mN. At temperatures below the PDA monolayer triple point of 17 °C, the L_1 state disappears and the gaseous state is found to coexist directly with the L_2 phase, which can then be compressed to the solid state. For longer chain length fatty acids, stronger van der Waals interchain attraction of the chain raises the triple point, and therefore, for example, compression of the gas phase of stearic acid (C_{18}) at 25 °C gives rise directly to the L_2 state.

Visual confirmation of two-phase coexistence and the first order nature of the G-LE, LE-L_2 and G-L_2 transitions of the medium chain length fatty acid amphiphiles has been obtained with fluorescence and Brewster angle microscopy (BAM). In fluorescence microscopy,[35-37] a small amount (< 1 mol %) of a water insoluble amphiphilic dye with a fluorescing chromophore is spread onto the air/water surface along with the amphiphile under study. The chromophore typically fluoresces in a lipid environment while being quenched by water. The monolayer is illuminated at the excitation frequency of the dye, and the interface fluorescence is filtered and observed. Different mechanisms produce the fluorescence contrast in the two main phase transitions, LE-LC and G-LE. In the dense monolayer phase transition, the dye fluoresces in the disordered, lipid-like LE phase while its heterocyclic head group forces its expulsion from the more tightly packed LC phase which appears dark. In the dilute monolayer phase transition, the dye is present in both surface phases, but its fluorescence is quenched in the gaseous phase where the isolated probe molecules sense only the aqueous environment. Thus the surface fluorescence shows the LE phase to be bright, while the G and LC phases appear dark. Using fluorescence microscopy, the G-LE and LE-L_2 phase coexistence states of myristic (C_{14}),[38] pentadecanoic,[34] and stearic acid[39] have been verified. In either phase transition, images showed many dark round domains, 10 - 100 µm in diameter, in a bright LE background. In the PDA LE-LC transition, the ratio of dark-to-light areas gave estimates for the pure phase densities consistent with surface pressure measurements.[34]

The origin of phase discrimination in Brewster angle microscopy[24, 25] is the reflectivity of the surface phases, as it is sensitive to refractive index and height differences. P-polarized laser light is incident at the Brewster angle for the clean water surface where the reflected light is minimized. The presence of a monolayer acts as a separate dielectric phase between the air and water, and increases the reflectivity in proportion to the surface density; coexistence between the G-LE, LE-L_2 and G-L_2 can be visualized since the various phases have different densities. Using BAM, the G-LE and LE-L_2 states of PDA[40, 41] and myristic acid,[24] and the G-LC coexistence of stearic acid,[41] have been studied, confirming the fluorescence results.

Direct evidence of the long range tilt ordering of the chains in the L_2 and more condensed liquid phases has also been obtained from polarized fluorescence microscopy (PFM) and polarization analysis of the reflectivity in BAM.[42, 43] With polarization and more intense illumination, liquid condensed domains, previously appearing uniformally dark (in fluorescence microscopy) or bright (in BAM), exhibit dim subdomains of different gray level fluorescence, determined by the orientation of the laser beam relative to the chain tilt angle in the subdomain. Textured images of stripe arrays in L_2 domains, in coexistence with the L_1 phase, have been observed with both techniques for myristic acid and PDA.[43-45] The disappearance or variation of these textures during isothermal surface pressure changes has been used to map other phase transitions between condensed states.[30, 46] The current understanding is that the phase diagram of the intermediate chain length amphiphiles has four states, as shown in Fig. 2.1. For longer chain lengths, a more complex phase diagram has been mapped.[28-30]

We might expect the phase behavior of bulk soluble surfactants to have features similar to those of the intermediate chain length bulk insoluble surfactants shown in Fig. 2.1. With surfactant solutions, surface pressures develop directly from the affinity of the dissolved surfactant to adsorb from the bulk to the surface. The added freedom of exchange

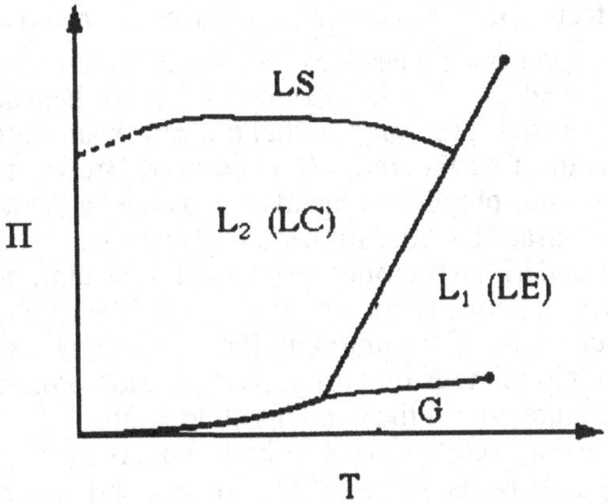

Fig. 2.1, The surface pressure/temperature phase diagram for intermediate chain length (C_{14}-C_{18}) insoluble surfactants at the air/water surface

with the subphase, along with temperature and dipolar replusion, balances the cohesive tendencies of van der Waals forces between chains to determine the monolayer state. In particular, exchange and the greater dipolar interactions make it more difficult to achieve the more condensed liquid and solid states, since soluble surfactants generally have larger polar groups than insolubles of the same chain length. Different adsorption states in monolayers at the air/water surface can in principle be inferred from the presence of cusps in the plot of the equilibrium surface tension γ as a function of the bulk surfactant concentration C with the slope magnitude $-d\gamma/dlnC$ showing a discontinuous rise at a particular concentration. From the Gibbs equation $d\gamma = -\Gamma RT\, dln\, C$, the slope of the γ-$ln\, C$ curve is proportional to the surface concentration Γ, where R is the universal gas constant and T is temperature. Therefore a true cusp indicates a first order phase transition from a less dense to a more condensed state.

Evidence for a gaseous / liquid expanded phase transition at room temperature from cusps in γ-$ln\, C$ isotherms at low surface pressure have been obtained by Matuura and coworkers[47-49] for a series of n-octyl (octanol, octanoic acid, sodium octanoate), n-dodecyl (sodium dodecylsulfate, dodecylammonium chloride, and dodecyltimethylammonium chloride) adducts, and an n-decyl amphiphile (decylammonium chloride). The cusps occur at tensions between 66.8 and 70.45 mN/m, and bulk concentrations well below the critical micelle concentration for each surfactant. From the slopes of the γ-$ln\, C$ curves, the surface densities at the discontinuities are in the range of 160 - 282 Å^2/molecule for the low density side, and 88 - 260 Å^2/molecule for the higher density side, directly correlated to the head group size and ionization. The large areas per molecule characterize this transition as gaseous to liquid expanded. Cusps in γ-$ln\, C$ isotherms have also been observed for n-octanol[50] and n-decanol[51].

The liquid expanded / liquid condensed phase transition so readily observed in the Π-A measurements for insoluble surfactants is not discernible in these γ-$ln\, c$ curves most likely because the density difference between these phases is relatively small ($\Delta A \sim 10$ Å^2/molecule), and therefore the slope change at the cusp is minute, making a cusp more difficult to distinguish. Hénon and Meunier studied monolayers of hexadecanoic acid adsorbed from aqueous solutions at pH 6.5 (providing partial ionization and limited solubility) onto a surface initially swept clean of surfactant.[52] They used Brewster angle microscopy to monitor the formation of surface phases during adsorption, and a Wilhelmy plate to measure concommittant changes in surface tension. Immediately after the surface was cleaned, BAM showed bright domains in a dark background growing at constant, near zero surface pressure indicating a first order transition from a gaseous state of zero reflectivity to a phase of higher surface density. Surfactant adsorbing onto the interface is taken up by the growing condensed phase at no decrease in tension during the transition. Polarization analysis of the reflected light exhibited texture within the condensed domains, indicating tilt ordering, and therefore a transition to the liquid condensed L_2 phase. This is the first conclusive evidence of the formation of the L_2 state by assembly from solution, and is consistent with the room temperature phase behavior of an insoluble monolayer of hexadecanoic acid at this temperature (at lower pH) which would also exhibit a gaseous / liquid condensed phase transition.

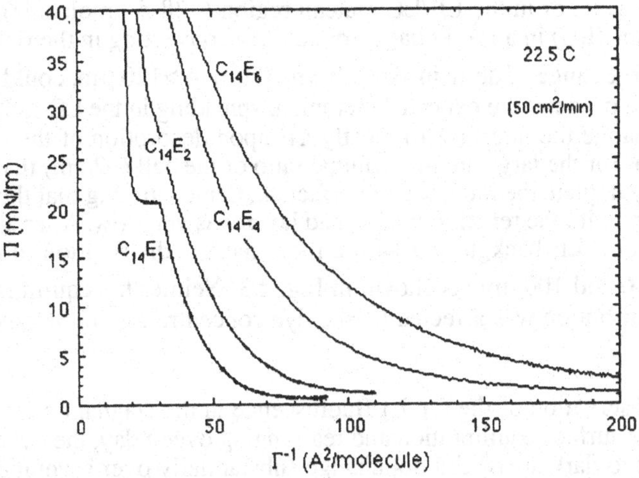

Fig.2.2, Surface pressure/area isotherms for a series of polyethoxylated surfactants with fixed chains, and head goups with increasing size. The temperature and compression rate are indicated in the figure

The above review of monolayer phase behavior of soluble surfactants indicates that these systems can exist in gaseous, liquid expanded, or liquid condensed states at room temperature, with shorter chains (octyl, decyl and dodecyl) showing a fairly universal gaseous to liquid expanded transition, and the longer chain partially ionized hexadecanoic acid having a transition from the gaseous to the liquid condensed state. For chain lengths less than 16, the possibility of the surfactant adsorbing to a liquid condensed state from the expanded one at higher bulk concentration cannot be evidenced from the γ-$ln\ C$ curves. In our own research, we have studied a series of polyethoxylated surfactants with a fixed chain length and a variable sized polar group, $C_{14}E_j$ ($CH_3(CH_2)_{13}(OCH_2CH_2)_j$-OH), j=2,4 and 6. We have used both surface pressure/area measurements and fluorescence microscopy to study the phase behavior. Figure 2.2 details the surface pressure measurements for these surfactants spread from a solvent directly on the surface. Since the polyethoxylates are soluble, the compression is done quickly to minimize desorption. For $C_{14}E_1$ we find that because the chain length is smaller than that of hexadecanoic acid, the monolayer is not as cohered, and a liquid expanded state exists and condenses to a liquid condensed state (LC), while for the two surfactants with the larger headgroup size, condensation is not possible and they remain in a liquid expanded state. The plateau density range, between 22 and 32 Å²/molecule, closely resembles that of PDA at this temperature, despite being at much higher surface pressure than the PDA transition (Π = 4.8 mN/m).[32, 34] The $C_{14}E_1$ LE and LC phase compressibilities are also comparable with those of PDA: χ_{LE} = 0.03 - 0.13 m/mN (0.08 m/mN for PDA) and χ_{LC} = 0.017 m/mN (0.01 m/mN for PDA). The presence of a gaseous/liquid expanded transition cannot be discerned from these measurements because the tension measurements are not sensitive in the range of less than one mN/m.

Further examination of the phase behavior of $C_{14}E_1$ using fluoresence microscopy was undertaken by spreading at known area per molecule $C_{14}E_1$ and the fluorescing dye NBD-HDA (10^{-2}-.5 mole percent probe to surfactant) on the surface of a stagnant cell. For a spread area per molecule of in the LE/LC plateau region (~28 Å^2/molecule), we observed circular dark domains (LC) in a bright background (LE) convecting in the field of view due to air currents. A large range of domain sizes, from 20 µm to >120 µm, could be seen in the 2 mm x 1.7 mm field of view. We expected that initial spreading in the LE-LC plateau region would over time change the interface to mostly LE upon desorption of the surfactant. We found, however, that for the large area-to-volume ratio of the cell (~2/cm) that the subphase did not substantially deplete the interface of surfactant, demonstrating that the bulk concentration in equilibrium with the remaining adsorbed layer was very low. After a day, the interface showed a uniform LE background with a more narrow distribution of domain sizes, typically between 50 and 100 µm, as shown in Fig. 2.3. Neither the equilibration time nor the domain size distribution was affected by the dye concentration in the monolayer, from 10^{-2} to 0.5 mol%.

The main observation of the $C_{14}E_1$ fluorescence in the stagnant cells, as alluded to above, is that despite surface equilibration and rearranging over a day, the overall proportion of LE to LC (bright to dark areas) did not change substantially over several days' observation. Little desorption occurred in our cells. Thus we could spread monolayers at different areas per molecule and qualitatively compare the proportion of LE to LC areas to our surface pressure isotherms. The proportion of bright LE regions rose as the average area per molecule was increased within the plateau. At areas beyond the surface pressure plateau, a bright LE phase was observed. Initial dye inhomogeneities with diffuse boundaries in the pure LE region between brighter and dimmer areas equilibrated in minutes to produce a uniformly bright LE phase throughout the entire interface. At much larger areas per molecule (> 400 Å^2/molecule) phase coexistence again appeared, providing evidence of the expected coexistence between gaseous and LE phases, with a corresponding tension indistinguishable from that of water. The domains relaxed to round shapes upon covering of the entire area, and after a day mostly small gas domains (< 20 µm) in an LE background were found.

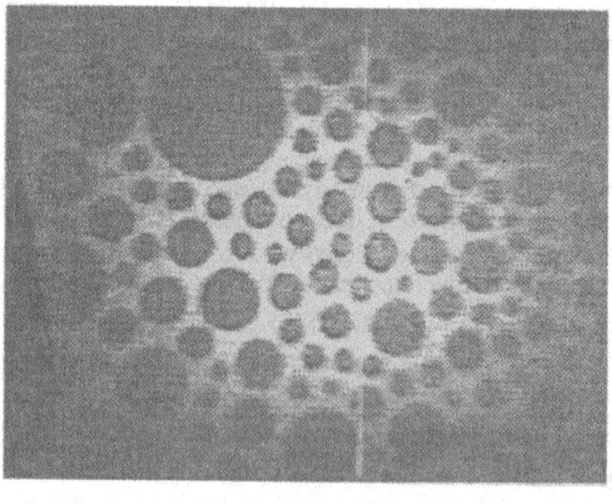

Fig. 2.3, Two phase (LE/LC) coexistence of C14E1 at 20 A^2/molecule: The LE phase containing the fluorescing dye is bright, while the LC phase in which the dye has been excluded because of the close packing of this phase, is dark.

3. MODELING THE SURFACTANT EQUATION OF STATE AND THE KINETIC EXCHANGE RATE

The evidence presented in Sec. 2 above indicates that for the more soluble surfactants, only the gaseous and liquid expanded phases are common, with the LC phase only occurring when the surfactant has very long chains (compared to the size of the headgroup) and is therefore of limited water solubility. In modeling equations of state and kinetic rates of exchange, account should necessarily be taken of the monolayer phase, with separate equations for the different phases. Most modeling efforts have not considered separately the G and LE phases; instead a single equation is written which is intended to be applicable for both states as a single phase. While this is certainly a valid approach above the critical point in temperature at which there is no distinction between the two phases, it is not accurate when coexistence is present. In what follows below, we will review the single phase models which are based on the Arrhenius rate law concept; modeling in which the gaseous and liquid expanded states are considered separately are given in Lin *et al.* [53]

The kinetic rate of adsorption of a surfactant molecule from the liquid sublayer immediately adjacent to an interface onto an interface is given in terms of a Arrhenius rate law equation in which the rates of adsorption and desorption are proportional to activation energies. For exchange between one sublayer and the surface, the rate Q is of the form:

$$Q = \beta C_s e^{-\frac{E_A}{RT}}(\Gamma_\infty - \Gamma) - \alpha e^{-\frac{E_D}{RT}}\Gamma \tag{3.1}$$

where Γ is the surface concentration, C_S is the sublayer concentration, α and β are kinetic rate constants for desorption and adsorption, respectively, E_A/RT and E_D/RT are, respectively, the activation energies for adsorption and desorption divided by the thermal energy, RT, and Γ_∞ is the maximum packing concentration of surfactant on the interface. This kinetic equation is developed in references [54-59]. From equation (3.1), the equilibrium adsorption Γ_o for a bulk concentration C_o is given by:

$$\frac{\Gamma_0}{\Gamma_\infty} = \frac{1}{1 + \dfrac{\alpha}{\beta C_0} e^{-(E_D - E_A)/RT}} \tag{3.2}$$

Equation (3.2) is the adsorption isotherm.

The activation energies can depend on the surface coverage Γ. The dependence of the activation energy for adsorption on the surface coverage describes the work necessary to insert a surfactant molecule into a monolayer already covered with concentration Γ. The dependence of the activation energy for desorption on the coverage describes the balance between the cohesive van der Waals energies between the nonpolar chains which tend to cohere the monolayer and increase the activation energy for desorption, and the repulsive interactions between the polar groups which tend to reduce the activation energy for desorption. When the activation energies are independent of the surface coverage, the adsorption isotherm is the Langmuir isotherm; in this case the exponential terms can be incorporated in the definitions of the rate constants α and β, and the Langmuir adsorption is simply written as:

$$\frac{\Gamma_0}{\Gamma_\infty}=\frac{1}{1+\frac{\alpha}{\beta C_0}} \tag{3.3}$$

If the dependence of the activation energies on the coverage is assumed linear, $E_A = E_A^o + \nu_A\Gamma$ and $E_D = E_D^o + \nu_D\Gamma$ then the adsorption isotherm becomes the Frumkin equation and is of the form:

$$\frac{\Gamma_0}{\Gamma_\infty}=\frac{1}{1+\frac{\alpha}{\beta C_0}e^{K\frac{\Gamma_0}{\Gamma_\infty}}} \tag{3.4}$$

where K is an interaction parameter, and is equal to $(\nu_A - \nu_D)\Gamma_\infty /RT$, and the constant adsorption energies are adsorbed into the rate constants.

The presence of cohesion or repulsion of the adsorbed molecules can markedly affect the adsorption isotherm. If we assume that the activation energy for adsorption is independent of the coverage, then Figure 3.1 contrasts the cases where cohesion dominates and the activation energy for desorption increases with coverage (K<0) and where repulsion is dominant and the activation energy for desorption decreases with coverage. (K>0). From the figure it is clear that for negative K the adsorption increases faster with the bulk concentration, as the presence of adsorbed molecules on the surface increases the desorption activation energies (and decreases the desorption rate constant), and allows more surfactant to adsorb in order to obtain equilibrium.

We have already mentioned that the adsorption of surfactant from the bulk water phase to the air surface or the oil/water interface lowers the free energy of the system. This energy reduction also reduces the surface or interfacial tension γ from the value of the clean

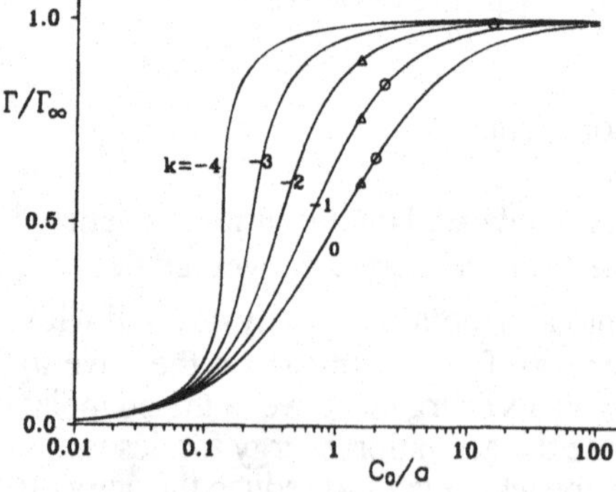

Fig. 3.1, Equilibrium adsorption (Γ/Γ_∞) as a function of nondimensional bulk concentration C_o/a where $a=\alpha/\beta$.

interface γ_c to a value which is a function of the surface concentration. At equilibrium and at constant temperature, the reduction in tension is given by the Gibbs equation:

$$d\gamma = -\Gamma RT \, d\ln C_o \qquad (3.5)$$

Equation 3.5 can be integrated by using the adsorption isotherms to obtain the equation of state which relates, at equilibrium the tension to the surface concentration ($\gamma(\Gamma)$). For example, for the Frumkin isotherm in which the activation energy for desorption depends linearly on the surface coverage, integration yields:

$$\gamma(\Gamma) = \gamma_c + RT\Gamma_\infty\left(\ln\left(1-\frac{\Gamma}{\Gamma_\infty}\right)-\frac{K}{2}\left(\frac{\Gamma}{\Gamma_\infty}\right)^2\right) \qquad (3.6)$$

where Π is the surface pressure. The equation of state for the Langmuir isotherm follows from 3.6 with K=0. A general analytical relationship for the equilibrium surface tension as a function of the bulk concentration C_o is in general not possible, but is obtained by numerically solving the equation of state and the adsorption isotherm (for example, for the Frumkin isotherm, eqs. 3.4 and 3.6). However for the Langmuir isotherm (K=0), an analytical relationship can be obtained, and is given as:

$$\gamma(\Gamma) = \gamma_c - RT\Gamma_\infty\left(\ln\left(1+\beta\frac{C_o}{\alpha}\right)\right) \qquad (3.7)$$

As we will describe in the next section, the most accessible experimental measurement is the equilibrium tension as a function of the bulk concentration of surfactant (C_o). From this measurement and the fitting to the theoretical relationship between the tension and the concentration, parameters of the adsorption model can be determined. For example, for Langmuir adsorption, the fitting of surface tension data to eq. 3.7 allows for the determination of the kinetic ratio α/β and the maximum packing concentration Γ_∞.

The effect of the interactions between the adsorbed molecules on the surface pressure Π (γ_c-$\gamma(\Gamma)$) is shown in Fig. 3.2 for the Frumkin isotherm where the surface pressure is plotted as a function of the area per molecule A(A=1/Γ) Note that when K is positive, repulsive interactions between the polar groups dominate, and for a given area per molecule the surface pressure is higher (the tension lower). For K negative cohesive interactions of the chains dominates, and there is a region where the surface pressure does not change significantly with a decrease in the area per molecule. In this region, the increased surface pressure de to the increased adsorption as the area per molecule is decreased is balanced by the condensing effect of the cohesive interaction. For even stronger interactions, the surface pressure can become exactly horizontal and not change with area per molecule for a region, in which case two phases exist on the surface as we discussed in the previous section.

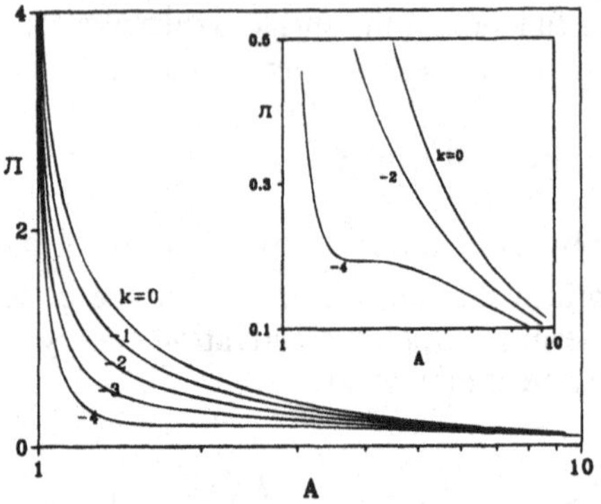

Fig. 3.2, Equilibrium surface pressure as a function of A, area per molecule (Γ_∞/Γ)

4. THE MEASUREMENT OF STATE EQUATION AND TRANSPORT PARAMETERS

The parameters in state equations are measured by fitting equilibrium surface tension/bulk concentration measurements as described in Sec. 3. The two most common and accurate methods for the measurements of the equilibrium air/water surface tension are the Wilhelmy plate technique, and the pendant bubble method. In the Wilhelmy place technique a thin platinum plate connected to a force transducer is placed on the air /water surface. Since water wets platinum, the water climbs the side of the plate to establish a zero contact angle in the liquid. The transducer measures the pull of the surface tension force on the plate, from which the tension is measured. In the pendant bubble technique, a pendant bubble is formed by the injection of gas through a needle immersed in the surfactant solution. Surfactant adsorbs onto the freshly formed interface, reducing the tension of the bubble surface and elongating the bubble until an equilibrium adsorption is obtained. Digitized video images of silhouettes of the bubble shape are captured, and the locus of points comprising the bubble interface is constructed from edge detection. The surface tension of the bubble interface is computed by matching the bubble profile to theoretical pendant shapes obtained by solving the Young-Laplace equation.

To illustrate the fitting of equations of state, in Fig. 4.1 we present some of our data for the equilibrium surface tension of the nonionic polyethoxylated surfactant $C_{12}E_6$ (the E indicates a $-OCH_2CH_2-$ ethoxy group and the ethoxy string terminates in an -OH) as a function of bulk concentration at the air/water interface using both the pendant bubble and the Wilhelmy plate techniques. We note that both techniques are in reasonable agreement. Also shown in the figure is the fitting of the data to the Langmuir and Frumkin equations of state. From the Langmuir relation we find $\Gamma_\infty=2.42\text{x}10^{-10}$mol/cm^2 and $\alpha/\beta=.0512$ mg/dm^3 and for the Frumkin, $\Gamma_\infty=3.48\text{x}10^{-10}$mol/cm^2, $\alpha/\beta=.0157$ mg/dm^3 and K= 6.652 (indicating

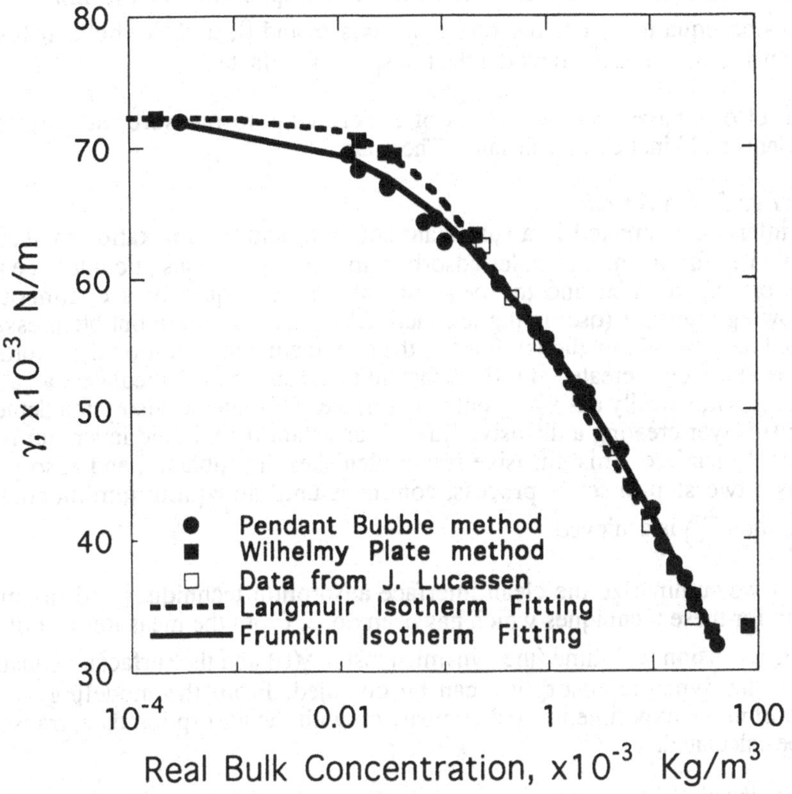

Fig. 4.1, Equilibrium surface tensions for air/$C_{12}E_6$ aqueous solutions and the Langmuir and Frumkin isotherm fitting.

repulsion dominance). We note finally from the figure that at a high enough concentration the tension does not change with an increase in concentration. This indicates the formation of surfactant aggregates in the bulk (micelles), and the concentration at which this occurs is the critical micelle concentration or CMC.

The rate of exchange of surfactant molecules between a fluid interface and the surrounding bulk phases is determined by the kinetic flux between the surface and the bulk sublayers next to the surface, and the diffusive and convective flux between these sublayers and the regions of the bulk phases extending away from the surface. The kinetic exchange between the sublayer and the surface has been described in Sec. **3** in terms of an Arrhenius rate law equation. Diffusive and convective fluxes are described by the convective diffusion equation. Accurate values for the transport coefficients appearing in the kinetic and convective diffusion equations, i.e. the rate constants α and β, and the bulk diffusion coefficient D are necessary in order to model the transport of surfactant.

Research efforts have used two types of experiments to measure the surfactant diffusion coefficients and kinetic rate constants. These are:

(i) *Clean Interface Adsorption*

A fresh interface is created in a surfactant solution, and the relaxation in surface tension is measured as surfactant molecules adsorb from solution onto a static interface (the Wilhelmy plate or ring method and the pendant bubble technique) or a continuously deforming or flowing interface (oscillating jet, inclined plane, maximum bubble pressure, growing drop, and drop weight methods). For $t<0$ the surfactant concentration is uniform in the bulk; when the interface is created at $t=0$, surfactant in the sublayer immediately adjacent to the interface begins kinetically adsorbing onto the surface. This kinetic adsorption depletes surfactant in the sublayer creating a diffusive flux of surfactant from liquid layers far from the interface onto the surface. This diffusive flux replenishes the sublayer, and adsorption and diffusion, as a two step in series process, continues until an equilibrium monolayer (surface concentration Γ_0) is achieved.

In Table I we summarize the clean interface adsorption techniques and the mass transfer modeling for these techniques which has been done. From the measurement of the relaxation in surface tension with time (the dynamic tension $\gamma(t)$) and the surfactant equation of state (Sec. **3**), the dynamic adsorption can be obtained. From the modeling of the surfactant transport in the experiments, and comparison with the adsorption data, transport parameters can be calculated.

(ii) *Re-equilibration*

An equilibrium monolayer is perturbed either with small amplitude periodic oscillations (Langmuir trough and oscillating bubble methods) or large amplitude area changes (elastic ring or float on the air/water surface of a Langmuir trough, a reversed funnel, pulsating bubble and shrinking or expanding pendant bubble techniques), and the tension oscillation or relaxation as the interface re-equilibrates is measured.

Again, by modeling the kinetic exchange and the bulk diffusive and convective transport with the convective-diffusion equation, $\Gamma(t)$ can be predicted as a function of the surfactant transport parameters and compared to the dynamic tension data to obtain transport parameters. The types of re-equilibration experiments are listed in Table II.

Table I
Clean Interface Adsorption Methods for Measuring Surfactant Exchange At the Air/Water Interface

Oscillating Jet	Surfactant adsorbs onto the surface of a jet issuing from an elliptical orifice; measurement of jet oscillation wavelength at different downstream positions provides surface tension as a function of surface age	One dimensional bulk diffusion and kinetic adsorption to a planar surface to describe transport to a moving surface element; convective transfer neglected.
Inclined Plane [60, 61]	Surfactant adsorbs onto the surface of a liquid film falling down an inclined plane; surface tension measured at downstream positions by Wilhelmy plate provides tension as a function of surface age.	
Max. Bubble Pressure [62-66] Growing Drop [67-71] Drop Weight Method [72-78]	Surfactant adsorbs onto a bubble growing into a liquid or a drop forming in air, both at the tip of a needle; pressure measured when hemisphere (maximum bubble pressure) or continuously (growing drop). Tension relaxation obtained from Young-Laplace relation Surfactant adsorbs onto drop detaching from capillary tip into air; weight of detached drops measured to obtain surface tension at instant of detachment	One dimensional bulk diffusion and kinetic adsorption to a stretching planar interface. Surfactant transport by convection accounted for in theory by assuming flow is purely radial. Diffusive transport assumed to occurs in a narrow layer around the bubble or drop.
Static Pendant Bubble or Drop [79-82] Wilhelmy Plate/ Ring [83]	Surfactant adsorbs onto a stationary, pendant bubble or drop rapidly formed at the tip of capillary; shape imaged and compared to solutions of Young-Laplace equation to obtain surface tension relaxation Surfactant solution placed in a open container, and Whilhelmy plate (ring) placed into surface. Surface swept clean of surfactant, surfactant adsorbs onto clean interface, and tension measured.	Radial diffusion and kinetic adsorption to a spherical bubble or drop (pendant technique); planar diffusion and kinetic adsorption to a flat surface (Whilhelmy plate or ring method). Convection assumed to have subsided in the measurement time period

Table II
Monolayer Perturbation Methods for Measuring
Surfactant Exchange At the Air/Water Interface

Small oscillatory area changes to a planar or spherical surface [84-88]	Two approaches: I. Area of a planar interface is perturbatively changed in a restricted domain such that the surfactant concentration remains spatially uniform; surface tension measured by Whilhelmy plate II. The surface of a spherical bubble or drop tethered to a needle is oscillated radially, and pressure inside bubble measured to obtain tension.	Sinusoidal variation in tension related to variation in surfactant concentration and to the diffusive and kinetic exchange assuming one dimensional diffusive transport and negligible convection.
Large area changes in a spherical surface [89-92]	Two approaches: I. Area of a pendant bubble cyclically changed, pressure in bubble measured to obtain tension. II. Jump change in area imposed on a pendant bubble, tension re-equilibration measured in static system by shape analysis.	One dimensional radial diffusion, kinetic exchange and (for cyclic changes) radial convection to the spherical surface
Small amplitude longitudinal or capillary waves [93-98]	In plane longitudinal waves created by sinusoidal motion of a wire on a planar interface, or transverse capillary waves created by electrocapillary excitation. Wavelength and damping coefficient measured from point of application of the waves	Coupled solutions to hydrodynamic surface wave and surfactant kinetic and diffusive mass transfer describe transport effects on wavelength and damping. Convection perturbatively small.
Large area changes in a planar surface [99-104]	Large amplitude, continual expansions or contractions of a planar surface created by stretching a floating elastic ring, moving a floating boom or adjusting the height of the free surface in a inverted funnel. Tension measured by Whilhelmy plate placed in surface.	One dimensional diffusion and kinetic exchange solved to determine tension as a function of time. Convective transport neglected.

In the above experiment methods summarized in Tables I and II, the interpretation of the surface tension relaxations begins by assuming that kinetic exchange is very fast, and that the transport is diffusion controlled over the time scales (or frequencies) of the experiment. With this assumption, an apparent diffusion coefficient is calculated. It is difficult to measure separately the diffusion coefficient of surfactant monomers by independent methods such as light scattering because the monomer concentration is usually very low. The apparent value is therefore usually compared to a value obtained from the Stokes-Einstein equation (assuming some value for the hydrodynamic radius based on molecular considerations); if the apparent value is in reasonable agreement with the Stokes-Einstein value then no further interpretation of the experiment is done, the apparent value is considered as the diffusion coefficient, and the kinetic constants remain unresolved. If the calculated apparent diffusion coefficient differs greatly from the Stokes-Einstein expectation, then the transport process is either modeled as kinetically controlled, and comparison with data yields the kinetic parameters, or as mixed, and kinetic constants and diffusion coefficients are derived from a fit of the data.

Most investigations have concluded that surfactant transport in the dynamic tension relaxation is diffusion controlled, and have computed diffusion coefficients. Results are listed in Table III. Marked kinetic effects have only been observed in clean interface adsorption for bolaform surfactants with two hydroxyl (1,2 and 1,8 thru 1,11 diols) or carboxcylic acid groups (1,6 thru 1,10 diacids), octanols of linear hydrocarbon chain with

Table III

Phosphene Oxides $DMC_{10}PO$, $DMC_{12}PO$ n-decyl or n-dodecyl dimethyl phosphene oxide	Drop Weight [77] Max Bubble Press. [62, 74] Wilhelmy Plate or Ring [83] Large Area Change [100, 105]
DEC_8PO, $DEC_{10}PO$, $DEC_{12}PO$ n-octyl, n-decyl or n-dodecyl diethyl phosphene oxide	Wilhelmy Plate or Ring [83] Drop Weight [77]
Polyethoxylated Alcohols $C_{12}E_6$, $C_{12}E_3$, $C_{14}E_6$	Max. Bubble Press [62, 74] Small Amplitude Osc. [106]
Triton X-100 $CH_3C(CH_2)_2CH_2C(CH_2)_2C_6H_6E_9$-OH	Static Pendant Bubble [80] Drop Weight Method [78]
Brij 58 ($CH_3(CH_2)_{15}E_9OH$)	Large Area Change [103]
Long chain n-alcohols n-octanol, n-nonanol and n-decanol	Oscillating Jet [107, 108] Maximum Bubble Pressure [109] Growing Drop [68] Static Pendant Bubble [53] Wilhelmy Plate or Ring [83] Large Area Change [101, 110]
Slightly Soluble Monolayers Lauric acid Myristic acid Dodecanol	Large Area Change [102, 108]
Alkanoic Acids and Salts Octanoic and Decanoic Acids	Small Amplitude Osc. [84] Longitudinal wave [111] Capillary Wave [95]
dodecyltrimethylamonium bromide	Inclined Plane [60, 61]
Sodium myristate	
Bolaform Surfactants 1,12 dodecanediol and diacid	Oscillating Jet [107] Large Area Change [102]

the hydroxyl at intermediate 2 thru 4 positions in the chain, and short and intermediate chain soluble n-alcohols (n=3-7). These results are listed in Table IV. For re-equilibration experiments, the only surfactants which have demonstrated kinetic effects are again a bolaform surfactant, 1,9 diol, and n-decanol (see Table IV). From the clean interface adsorption and re-equilibration experiments, the values determined for α are uniformly of the order of

10^2 sec^{-1}, and values for ß follow from the equilibrium ratio. Presumably, the slow kinetic exchange of the bifunctional diols, diacids and 2,3,4 octanols arises because the exchange of a surfactant with two separated polar groups or a polar group in an intermediate position in a linear hydrocarbon chain is a slow process. This point of view is supported by the fact that as the intervening chain length increases for the diols and diacids, kinetic effects are no longer present and the process becomes diffusion controlled. Kinetic effects in the exchange of n-octanol and n-decanol from equilibrium monolayers are probably due to strong cohesive interactions between the hydrocarbon chains of these surfactants, facilitated by their slender architecture which allows them to pack lengthwise closely together: This cohesion reduces the desorption kinetic rate by increasing the activation energy for desorption, E_D. This reduction in the desorption rate is more effective the larger the surface density, and is the reason why adsorption onto a clean interface of n-decanol can be observed to be diffusion limited, while re-equilibration or oscillation of an equilibrium layer is mixed.

Table IV

Surfactants Which Have Demonstrated Kinetic Effects

Alcohols: n-propanol, n-butanol and n-pentanol ,n-hexanol, n-heptanol	Oscillating Jet [112-114] Maximum Bubble Pressure [109]
Branched Octanols (2,3,4 octanol)	Oscillating Jet [107]
Decanol	Jump area change in pendant bubble [92]
Bolaform Surfactants: 1,2 and 1,8 - 1,11 n-diols and 1,6 - 1,10 n-diacids	Oscillating Jet [107, 115] Large area change[99]

Aside from the n- and branched alcohol and bifunctional diacid and diol studies, no other kinetic measurements exist. The reason usually given as to why kinetic effects are not manifest is that kinetic exchange timescale is too short. In both clean interface adsorption and re-equilibration, the kinetic step precedes and initiates the bulk diffusive process, i.e. surfactant first kinetically adsorbs onto a clean interface, reducing the sublayer concentration which drives the diffusion towards the surface, or desorbs off of the overcrowded interface, increasing the sublayer concentration which drives the diffusion away from the surface. Consequently the time dependent change in surface concentration of surfactant is controlled by kinetics for short times and diffusion for longer times. If the kinetic exchange is too fast, then measurements only capture the diffusion limited regime.

Our own recent research has shown that kinetic rate constants can be measured if the bulk concentration is large enough for clean interface adsorption, and the interface compression is large enough for re-equilibration. We study the adsorption of the polyethoxylated surfactant $C_{12}E_6$ at the air/water interface using the pendant bubble technique. The equation of state for this surfactant has already been given above from the equilibrium measurement of the tension against concentration, and we use here the Frumkin model. The pendant bubble technique, as summarized in Table I, is based on the analysis of the shape of pendant bubbles formed and tethered to the tip of a needle placed in the surfactant solution. We use the technique for both clean interface adsorption and re-equilibration studies . For clean interface adsorption, a pendant bubble is formed quickly (<1sec.) by the injection of gas through a needle immersed in the surfactant solution. Surfactant adsorbs onto the freshly formed interface, reducing the tension of the bubble surface and elongating the bubble. Digitized video images of the bubble elongation are captured sequentially in time; for each digitized image, the locus of points comprising the bubble interface is constructed, and the surface tension of the bubble interface at that time is computed by matching the bubble profile to theoretical pendant shapes obtained by solving the Young-Laplace equation. Thus from the sequence of images, a surface tension relaxation can be constructed. Re-equilibration experiments are undertaken by first allowing a bubble to form an equilibrium monolayer on its surface, and then quickly reducing the bubble area (<1sec). Video capture of the bubble shape as the surfactant re-equilibrates provides the surface tension re-equilibration. For both clean interface adsorption and re-equilibration, bubble shapes are captured only after the bubble is created, or the jump change in area is complete, so that convective effects are not important in the surfactant mass transfer. We develop solutions of the kinetic-diffusion equations and compare to the experimental dynamic tensions. Comparisons at increasing concentrations (for clean interface adsorption) or larger initial crowding (for re-equilibration), show the progressive effect of kinetics, and we calculate from this shifting in control of the transport the kinetic rate constants for $C_{12}E_6$.

In Figs. 4.2 and 4.3 are presented clean interface adsorption and re-equilibration experiments. The clean interface adsorption experiments are undertaken at five increasing bulk concentrations (a-e, as noted in the caption to the figure). The re-equilibrations are done at two bulk concentrations, and each concentration at two different compression ratios (A_f/A_i) as noted in the figure caption. We describe next the transport model to the pendant bubble interface.

The bubble is assumed to be a sphere of radius R_O immersed in an infinite continuous liquid phase. A spherical coordinate system with radial coordinate r and an origin at the bubble center is used. We assume the diffusive transport to be spherically symmetric so that the bulk concentration is only a function of r and t (C(r,t)), and the surface concentration only a function of t ($\Gamma(t)$). For clean interface adsorption, at time t=0, the point in time experimentally at which the bubble and the surrounding liquid phase have come to rest, we assume the surface concentration is equal to zero, and the bulk concentration is uniform. For re-equilibration, at time t=0, we calculate the compressed initial density (Γ_i) from the surface tension immediately after compression using the equation of state. In modelling the kinetic transport, we will only consider the activation energy for desoption to be a function of the surface concentration (ν_A=0); this reduces the number of parameters to be fit to just two since ν_D is given by K.

Nondimensionalizing the bulk concentration by a, the ratio of α/β, ($C^* = C/a$), the surface concentration by Γ_∞ ($\Theta = \Gamma/\Gamma_\infty$), time by the scale for bulk diffusion $(\Gamma_\infty/a)^2/D$ ($\tau = tD(a/\Gamma_\infty)^2$), and the radial coordinate by Γ_∞/a ($r^* = r(a/\Gamma_\infty)$), the nondimensional form of the kinetic equation is:

$$\frac{d\theta}{d\tau} = \lambda\left(C_s^*(1-\theta) - \theta e^{K\theta}\right) \qquad (4.1)$$

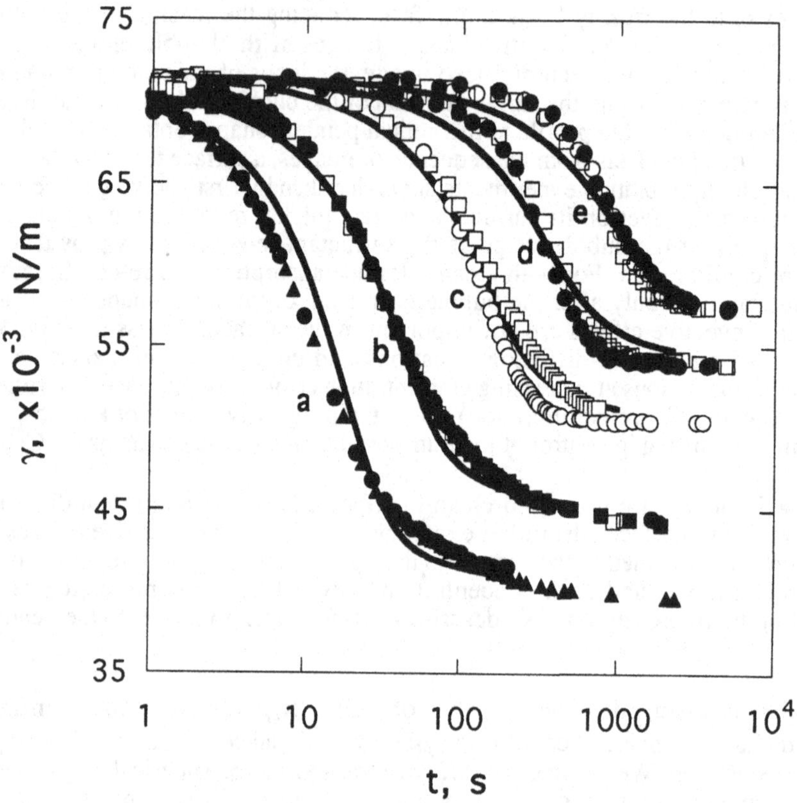

Fig.4.2, Dynamic surface tensions of clean surface \adsorption in $C_{12}E_6$ aqueous solution and mixed diffusive-kinetic control theoretical fit with $D=6.0\cdot10^{-10}$ m^2/s and $\beta=4.0$ $m^3/(mol\ s)$: a--$C_0=12.28\cdot10^{-3}$ kg/m^3; b--$C_0 = 6.14\cdot10^{-3}$ kg/m^3; c--$C_0=2.456\cdot10^{-3}$ kg/m^3; d--$C_0=1.228\cdot10^{-3}$ kg/m^3; e--$C_0=0.614\cdot10^{-3}$ kg/m^3

where $K= -\nu_D\Gamma_\infty /RT$, $C_S{}^*$ is the nondimensional sublayer concentration ($C_S{}^* = C^*(r^*=b,t)$), b is the ratio of the bubble radius to the length scale Γ_∞/a, and λ is the ratio of the bulk diffusion timescale $((\Gamma_\infty/a)^2/D)$ to the kinetic scale for desorption $(\alpha)^{-1}$:

$$\lambda=\alpha(\Gamma_\infty/a)^2/D) \qquad (4.2)$$

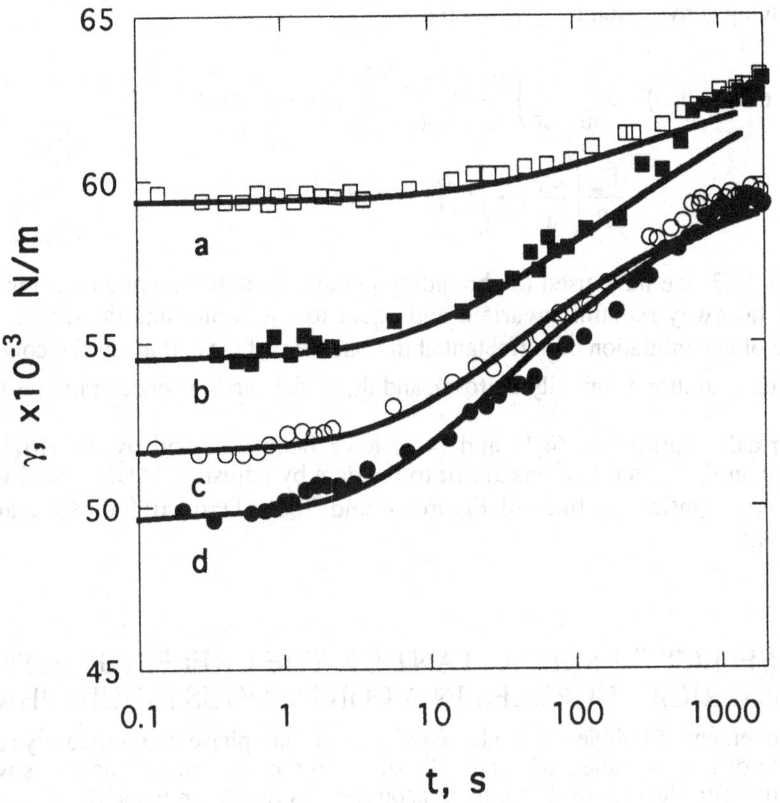

Fig.4.3, Dynamic surface tensions of re-equilibration in $C_{12}E_6$ aqueous solution and mixed diffusive-kinetic control theoretical fit with $D=6.0\cdot10^{-10}$ m^2/s and $\beta=4.0$ $m^3/(mol\ s)$: a--$C_0=0.18\cdot10^{-3}$ kg/m^3 and $A_f/A_i=0.77$; b--$C_0 = 0.18\cdot10^{-3}$ kg/m^3 and $A_f/A_i=0.59$; c--$C_0=0.40\cdot10^{-3}$ kg/m^3 and $A_f/A_i=0.66$; d--$C_0=0.40\cdot10^{-3}$ kg/m^3 and $A_f/A_i=0.60$.

In order to solve eq. (4.1), the sublayer concentration must be known. This concentration is obtained by solving for the diffusive transport in the bulk. Diffusive transport to the static bubble surface is described by Fick's law which can be integrated to yield the following convolution integral in time:

$$\theta(\tau)-\theta(0)=2\frac{C_0}{a}\left(\frac{\tau}{\pi}\right)^{1/2}-2\left(\frac{1}{\pi}\right)^{1/2}\int_0^{\tau} C_s^*(\tau-\tau')\,d\sqrt{\tau'}$$
$$-\frac{\Gamma_\infty}{aR_0}\left(\frac{C_0}{a}\tau-\int_0^{\tau} C_s^*(\tau-\tau')\,d\tau'\right) \tag{4.3}$$

In deriving eq. (4.3) we have used the boundary conditions that the nondimensional bulk concentration far away remains invariant and equal to C_0/a, and that the diffusive flux equals the rate of accumulation of surfactant at the bubble surface and the initial conditions. that the bulk concentration is initially uniform, and the initial surface concentration is $\theta(0)$.

Numerical solutions of (4.1) and (4.3) have been obtained by discritizing the convolution integral. The simulations are fit to the data by adjusting D and ß, and we find excellent fits (the continuous lines of Figures 4 and 5) for $D=6.x10^{-6}$ cm^2/sec and ß = 4. m^3/(mole sec).

5. THE EFFECT OF SURFACTANT ON THE CREEPING MOTION OF SPHERICAL BUBBLES IN A CONTINUOUS LIQUID PHASE

The movement of bubbles through a continuous liquid phase is significantly reduced by the presence of bulk soluble surfactants dissolved in the continuous phase. As we first mentioned in the Introduction (Sec. **1**), the reason for this reduction is the development of surface tension gradients (Marangoni forces) on the bubble surface which retard the interfacial mobility. When a freshly created bubble begins to move in a surfactant solution, surfactant adsorbs kinetically onto the surface of the fluid particle from the liquid sublayer immediately adjacent to the interface. The adsorbed surfactant is convected by the surface flow to the particle's trailing end. Accumulation at this end causes kinetic desorption into the bulk sublayer, and the sublayer concentration at the back increases above the value far from the interface. This difference gives rise to a diffusive flux away from the trailing end. Similarly at the front end kinetic adsorption occurs from the sublayer since the front surface is swept clean of surfactant. The sublayer concentration adjacent to the leading end of the particle decreases creating a bulk diffusive flux from the bulk to the front end. Eventually a steady state develops: In this state, the surface concentration at the back end has increased to the point where the desorption rate, proportional to the difference between the surface and sublayer concentration, balances the convective rate. In addition, the sublayer concentration has increased sufficiently so that the diffusive flux away from the particle surface, proportional to the difference between the sublayer and far field concentration, balances the kinetic desorption. At the front end, the surface concentration becomes reduced enough so that kinetic adsorption balances convection, and the sublayer concentration becomes reduced enough so that diffusion to the surface balances adsorption. Consequently, in this steady state the concentration is considerably higher at the rear than at the front of the particle, and the interfacial tension is lower at the back relative to the front. This interfacial tension difference creates a Marangoni stress along the surface as the front end tugs at the rear. The direction of this surface stress is opposite to that of the surface flow, and thus the adsorption of surfactant onto the particle interface acts to reduce the surface flow and hinder the interfacial mobility. The less mobile an interface, the more drag is exerted by the continuous phase on the particle as it moves through the medium, and the smaller is the velocity of the particle for a constant driving force.

In this section we show how these Marangoni forces are coupled to the hydrodynamic equations, and we obtain expressions for the terminal velocity as a function of the bulk concentration of surfactant, and the surfactant transport parameters. We choose the simplest case of a bubble rising by buoyancy, and assume the flow to be axisymmetric, and the bubble velocity to be small enough that inertial forces can be neglected with regard to viscous forces (Stokes or inertialess flow), and that capillary forces are stronger than either viscous or inertial forces, so that the bubble remains a section of a sphere. Far from the moving bubble, the bulk concentration is C_o and this is assumed to be less than the critical micelle concentration so that only monomer transport is present in the bulk.

We begin the theoretical formulation by discussing the hydrodynamic equations for the fluid flow in the continuous phase due to the bubble rise. The article by Levan [116, 117] provide a review to this formulation. We will describe the flow in a frame in which the particle is stationary, and the flow at infinity is uniform. We will use nondimensional variables throughout; lengths are scaled by the particle radius a, and velocity by the Hadamard-Rybcyznski value U_{HR} (= $\rho g a^2/3\mu$ where μ and ρ are the viscosity and density, respectively, of the continuous phase). A spherical coordinate system (r, θ) is located with its center coincident with the bubble center; the spherical angle θ of this coordinate system is measured from a z axis passing through the bubble center and aligned with the direction of the uniform flow. (This angle is measured from the leading edge of the bubble.) As the Reynolds number is assumed to be small, the flow in the continuous phase is inertialess. If we also assume that the flow is axisymmetric, then the Stokes equations can be written in terms of a (dimensionless) stream function $\psi(\theta,r)$ (nondimensionalized by a^2U_{HR}) by the equation $E^2(E^2\psi) = 0$, where E^2 denotes the axisymmetric stream function operator and the r and θ components of the nondimensional velocity (V_r and V_θ, respectively, nondimensionalized by U_{HR}) are defined as:

$$V_r = -\frac{1}{r^2 \sin\theta}\frac{\partial\psi}{\partial\theta} \qquad V_\theta = \frac{1}{r\sin\theta}\frac{\partial\psi}{\partial r} \tag{5.1}$$

A general solution for the stream function equation can be written as an infinite series of Gegenbauer polynomials $C_n^{-1/2}(\theta)$. The correct series solution which matches the uniform flow at infinity and a zero normal velocity at the bubble surface is of the form:

$$\psi(r,\theta) = U(r^2 - r)C_2^{-1/2}(\theta) + \sum_{n=2}^{\infty} B_n(r^{-n+1} - r^{-n+3})C_n^{-1/2}(\theta) \quad (5.2)$$

where U is the (nondimensional) terminal velocity (nondimensionalized by U_{HR}) and is as yet unknown, and B_n are undetermined coefficients.

The set of constants B_n can be obtained by satisfying the tangential stress boundary condition on the particle surface; this relation balances the viscous stress exerted by the continuous phase on the surface with the surface tension gradient due to the surfactant adsorption. In terms of the stream function this equation is of the form:

$$\psi_{rr} - 2\psi_r = \frac{1}{\mu U_{HR}} \frac{\partial \gamma}{\partial \theta} \quad (5.3)$$

where γ denotes the *dimensional* surface tension, the subscript indicates partial differentiation with respect to r, and the equation is evaluated at the particle surface.

We shall show below how the surface tension distribution is obtained by the solution of the surfactant transport equations. The final unknown, the terminal velocity U, is obtained by an overall force balance on the bubble surface which equates the drag exerted by the continuous phase on the bubble to the buoyant force $4\pi\rho g a^3/3$. From this relation it can easily be shown that the terminal velocity is obtained from B_2.

The surface tension distribution is calculated by solving the surfactant concentration distribution on the bubble surface. This concentration is obtained from the surface mass balance which equates the kinetic flux onto the surface (Q, as discussed in Sec. 3), to the convective transport along the surface.

$$\frac{1}{\sin\theta} \frac{\partial}{\partial \theta}[\sin\theta \cdot \Gamma V_\theta(r=1,\theta)] = \frac{aQ}{\Gamma_{eq} U_{HR}} \quad (5.4)$$

where Q is the *dimensional* kinetic flux, and Γ is the nondimensional surface concentration (nondimensionalized by the equilibrium concentration, Γ_o). In formulating the above, surface diffusion is neglected. We will use the Langmuir framework as described in Sec. III to describe the kinetic exchange. Thus

$$\frac{aQ}{\Gamma_{eq} U_{HR}} = Bi\left[kC_s(\frac{1}{Y} - \Gamma) - \Gamma\right] \quad (5.5)$$

where $Y=\Gamma_o/\Gamma_\infty$, C_S is the sublayer concentration, Bi (the Biot number) is a measure of the ratio of the desorption rate to the convection rate

$$Bi = \frac{\alpha a}{U_{HR}}$$

and k is the nondimensional bulk concentration:

$$k = \frac{\beta C_o}{\alpha}$$

$$Y = \frac{\Gamma_o}{\Gamma_\infty} = \frac{k}{k+1}$$

Using the Langmuir equation of state, the stress balance at the interface, written in terms of the surface concentration, becomes:

$$\psi_{rr} - 2\psi_r = Ma \cdot Y\left[\frac{1}{1-Y\Gamma}\right]\frac{\partial\Gamma}{\partial\theta}$$
$$Ma = \frac{RT\Gamma_\infty}{\mu U_{HR}} \qquad (5.6)$$

In the above, Ma is the Marangoni number, and represents the ratio of interfacial tension gradient forces to viscous forces. To complete the formulation, the sublayer concentration must be specified. This concentration is determined from the solution of the convective mass transfer equation in the bulk, subject to matching the (nondimensional) concentration field $C(r,\theta)$ (nondimensionalized by C_o) to the uniform value at infinity, and to matching of the diffusive flux onto the surface to the kinetic flux:

$$Bi\left[kC_s\left(\frac{1}{Y} - \Gamma\right) - \Gamma\right] = \frac{1}{\Phi}\frac{\partial C}{\partial r}_{r=1}$$
$$\Phi = Pe\frac{\Gamma_o}{aC_o} \qquad (5.7)$$

In the above, Pe is the Peclet number for bulk mass transfer (Pe=Ua/D; D is the monomer bulk diffusion coefficient.)

Most research has studied the case in which bulk diffusion is much faster than the kinetic step ($\Phi Bi \ll 1$). In this case, the bulk concentration field is uniform ($C_s=1$), and the surfactant distribution is determined soley by the kinetic step. We will confine our discussion to the kinetic regime; the effect of bulk diffusion is given in the articles of Levan and Newman [118] and Levan and Holbrook [119, 120].

For the regime of kinetic transport, when Bi becomes very small ($Bi \to 0$), convection of surfactant to the trailing pole is not compensated for by desorption. The velocity (but not the surface concentration) becomes zero in a stagnant cap ($V_\theta(r=1,\theta)=0$) at this pole in order to make the convective flux equal to zero, while the surface concentration (but not the tangential surface velocity) is zero ($\Gamma(\theta)=0$) in the region on the surface forward of the stagnation zone so that no surfactant is convected to the back. This regime in which surfactant collects in a cap has been studied by Davis and Acrivos [121], Sadhal and

Johnson[122] and He *et al* [20]. Mathematically the stagnant cap solution arises from the fact that in the limit as $Bi \to 0$, the surface conservation equation reduces to

$$\frac{1}{\sin\theta}\frac{\partial}{\partial\theta}[\sin\theta \cdot \Gamma V_\theta(r=1,\theta)] = 0 \tag{5.8}$$

which has a solution in which $V_\theta(r=1,\theta)=0$ for $\phi<\theta<\pi$, and $\Gamma(\theta)=0$ for $0<\theta<\phi$ where ϕ is the cap angle. An exact solution for the (nondimensional) terminal velocity U in terms of the cap angle has been developed by Sadhal and Johnson [122] by solving the mixed boundary value problem posed by the stagnant cap solution together with the Gegenbauer expansion. Their result for $U(\varphi)$ is

$$U = \frac{2\pi}{\left[\phi + \frac{1}{2}\sin\phi - \frac{1}{2}\sin 2\phi - \frac{1}{6}\sin 3\phi + 2\pi\right]} \tag{5.9}$$

The cap angle is left unspecified, and must be determined by first computing the surfactant distribution in the cap from the tangential stress condition evaluated as a function of the cap angle. From this distribution, the total amount of surfactant adsorbed can be obtained. A second relation for the total amount adsorbed can be obtained by integrating the surface mass balance equation over the cap. By equating the two relations the cap angle as a function of the Marangoni number (Ma) and the nondimensional bulk concentration (k) can be obtained. We have used this procedure for the case of Langmuir adsorption (He *et al.* [123]) for which the total (dimensional) amount adsorbed (M) can be shown from the conservation equation to be : $M=4\pi a^2\Gamma_o= 4\pi a^2\Gamma_\infty k/(1+k)$. Consider first the case in which Ma is very large, and tension gradient forces are much larger than the viscous forces which tend to compress the cap. As a result, the adsorbed monolayer cannot be compressed very much by viscous forces, and the cap angle increases rapidly with k. In Fig. 5.1 we plot $k(\phi)$ for Ma=10, 10^2 and 10^3, and we note that the interface becomes completely immobile ($\phi=\pi$) for k<.1 Thus when Ma is large, very little bulk concentration of surfactant is necessary to cause the bubbles to move as solid spheres. Alternatively, as Ma decreases, the compressibility of the monolayer increases, and larger bulk concentrations (larger k) are necessary to completely immobilize the surface. In Fig. 5.2 the cap angle is computed as a function of k for Ma=.1, .5 and 1, and we see for the smallest value of Ma (.1), k must be larger than 10 to completely immobilize the surface. We also compare in figures 5.1 and 5.2 the cap angles which are obtained with a linear (or Henry's law) kinetic relation with those obtained using the Langmuir isotherm. Linear kinetics has often been used to describe surfactant adsorption in the modelling of interfacial flows with surfactant (see, for example, Levan and Holbrook [119-120]). In the linear regime, the surface concentration is assumed to be much less than the maximum packing concentration ($\Gamma \ll \Gamma_\infty$), and the kinetic rate, adsorption isotherm and equation of state are given by:

$$Q = \beta C_s - \alpha\Gamma$$

$$\frac{\Gamma_0}{\Gamma_\infty} = k \tag{5.10}$$

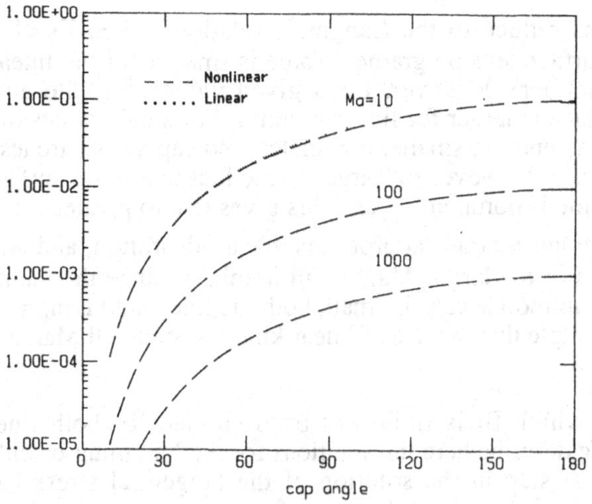

Fig. 5.1 Large Marangoni number solutions for the cap angle ϕ as a function of the nondimensional bulk concentration k. The dashed line is the numerical solution, while the dotted line is the solution for linear kinetics.

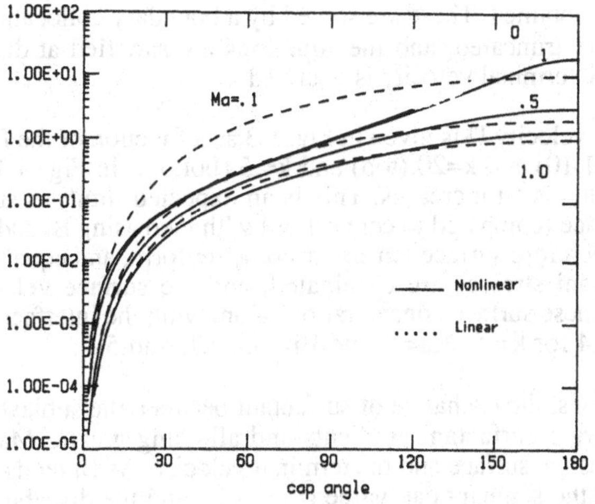

Fig. 5.2, Small Marangoni number solutions for the cap angle ϕ as a function of the nondimensional bulk concentration k. The solid line is the numerical solution, while the dashed line is the solution for linear kinetics.

$$\gamma(\Gamma) = \gamma_c + RT\Gamma$$

The above equations reduce to the Langmuir relations when $\Gamma << \Gamma_\infty$. For any surface concentration, the surface tension gradient force is smaller for the linear relation (with the same Marangoni number); however, for a given value of k, the amount of surfactant adsorbed on the surface is larger for linear kinetics. For small values of k, the fact that the interfacial tension gradients are smaller dominates, and cap angles are less for the linear case than the Langmuir case. However for larger k, the fact that more surfactant is adsorbed in the Henry's law regime is dominating, and this gives rise to greater cap angles. In the limit in which $k \to 0$, Langmuir kinetics approaches linear adsorption, and both approaches give the same cap angle. For the larger Marangoni numbers, since the value of k necessary to reach the completely immobile case is small, both the linear and Langmuir cap angles agree. We see from this example that the use of linear kinetics for small Marangoni numbers is not justified.

The case in which Bi is finite has been studied for both linear, Langmuir and Frumkin kinetics. We describe here the solutions for the Langmuir calculations of Chen and Stebe [117]. The first step in the solution of the tangential stress balance and surface conservation equation is to expand the surface concentration in a Legendre series, $\Gamma(\theta) = \sum_{n=0}^{\infty} a_n P_n(\theta)$ where $P_n(\theta)$ are Legendre polynomials. When this expansion is substituted into the stress balance and conservation equation, a set of two equations for the two unknown sets of coefficients, the surface concentration coefficients (a_n) and the velocity coefficients (B_n) are obtained. These are solved by a boundary collocation method, in which the infinite series are truncated, and the equations are satisfied at discrete points on the surface. From B_2, the terminal velocity is obtained.

The terminal velocity U is given in Fig. 5.3 as a function of the Biot number for four values of Ma (.1,.5,1,10) and k=20 (top) and k=.5 (bottom) in Fig. 5.3 For both values of k, U approaches unity as Bi increases. This is an expected limit; as surfactant exchanges rapidly with the surface (compared to convection) with increasing Bi and fixed k, the surface concentration and therefore surface tension become uniform, and equal to their equilibrium values. All Marangoni stresses are eliminated, and the surface velocity approaches its unretarded profile. These surface concentrations along with the interfacial tension gradients, are shown in Fig. 5.4 for k=.5, Ma=.5 and Bi=.05,.5,1, and 50.

As Bi decreases, the exchange of surfactant between the sublayer and the surface is reduced, creating larger surfactant gradients and allowing greater Marangoni stresses to develop. This reduces the surface and the terminal velocity. As Bi tends to zero the terminal velocity approaches the stagnant cap value (Fig. 5.2), and the distribution approaches the stagnant cap distribution as is evident by the Bi=.05 curve in Fig. 5.4. From Figs. 5.1 and 5.2, for k=20, and Ma=.1-10, the bulk concentration is high enough so that the cap angle is π. Thus in Fig. 5.3 when Bi tends to zero for k=20 and Ma = .1-10, the (nondimensional) terminal velocity is 2/3 (note the intercepts in Fig. 5.3). For k=.5, only when Ma=10 is the bulk concentration large enough so that the cap spans the entire bubble surface (Fig. 5.1), and U=2/3 (see the intercept in 5.3 for Ma=10). For $.1<Ma<1$ and k=.5 in Fig. 5.2, the cap angles are less than π and the terminal velocities from Fig. 5.3 are greater than 2/3.

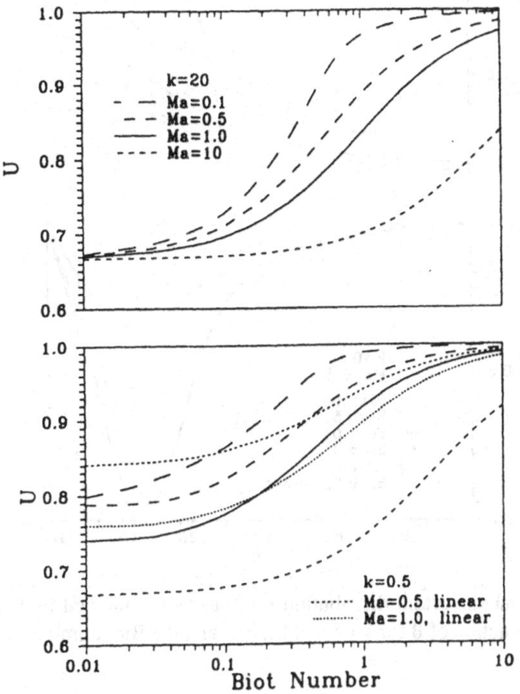

Fig. 5.3, The terminal velocity as a function of the Biot number for k=20 (top) and k=.5 (bottom) and for the Marangoni numbers indicated.

In Fig. 5.3 we also present results for the terminal velocity as a function of the Biot number for linear kinetics, for k=.5 and Ma=.5 and 1. Departures from the linear limit are already apparent at this value of k, and the stagnant cap asymptotes for the velocities are greater than those for Langmuir kinetics, indicating smaller cap angles, in agreement with Fig. 5.2 for k=.5 and Ma=.5 and 1. In Fig. 5.4, the surfactant distribution for linear kinetics is plotted for Bi=.5 (k=.5 and Ma=.5); we see that the linear equation of state allows the monolayer to be compressed above the maximum packing concentration.

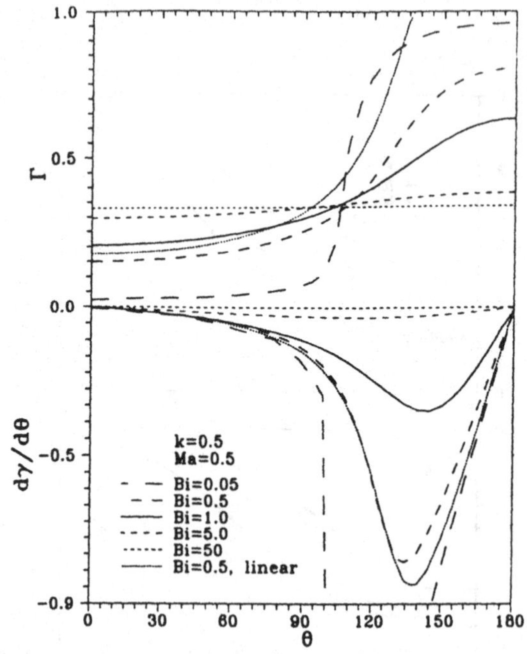

Fig. 5.4, The nondimensional surfactant distribution Γ (nondimensionalized by Γ_∞) and surface tension gradient as a function of θ for k=.5 and Ma=.5, and the Biot numbers indicated.

REFERENCES

1. M. Rosen: Surfactants and Interfacial Phenomena, John Wiley, New York, 1989.
2. C.-H. Chang and E. Franses: Colloids and Surfaces A 100 (1995) 1-45.
3. D. Edwards, H. Brenner and D. T. Wasan: Interfacial Transport Processes, Butterworth-Heineman, 1991.
4. J.C. Slattery: Interfacial Transport Phenomena, Springer Verlag, 1990.
5. I. Ivanov: Thin Liquid Films, Marcel Dekker, New York, 1988.
6. J. C. Lee and T. D. Hodgson: Chemical Engineering Science 23 (1968) 1375.
7. T. D. Hodgson and J. C. Lee: Journal of Colloid and Interface Science 30 (1969) 94.
8. B. P. Radoev, D. S. Dimitrov and I. B. Ivanov: Colloid and Polymer Science 252 (1974) 50.
9. I. Ivanov and D. S. Dimitrov: Colloid and Polymer Science 252 (1974) 982.
10. A. D. Barber and S. Hartland: Canadian Journal of Chemical Engineering 54 (1976) 279.

11. T. T. Traykov and I. Ivanov: International Journal of Multiphase Flow 3 (1977) 471.
12. Z. Zapryanov, A. K. Malhotra, N. Aderangi and D. T. Wasan: International Journal of Multiphase Flow 9 (1983) 105.
13. A. K. Malhotra and D. T. Wasan: Chemical Engineering Communications 55 (1987) 95-128.
14. W. J. Milliken, H. A. Stone and L. G. L. A: Phys. Fluids A 5 (1993) 69-79.
15. W. Milliken and L. G. Leal: J Colloid and Interface Science 166 (1994) 275-285.
16. S. A. K. Jeelani and S. Hartland: Journal of Colloid and Interface Science 164 (1994) 296-308.
17. G. Hirasaki and J. P. Lawson: SPE Journal 25 (1986) 176.
18. G. M. Ginley and C. J. Radke: ACS SYmposium Series 396 (1989) 480.
19. A. Borhan and C. Mao: Phys. Fluids A 4 (1992) 2628-2640.
20. Z. He, Z. Dagan and C. Maldarelli: Journal of Fluid Mechanics 222 (1991) 1-32.
21. J. Ratulowski and H. C. Chang: Journal of Fluid Mechanics 210 (1990) 303.
22. G. L. Gaines: Insoluble monolayers at liquid-gas interfaces, Wiley-Interscience, New York, 1966.
23. K. J. Stine: Micros. Res. Tech. 27 (1994) 439-450.
24. S. Hénon and J. Meunier: J. Chem. Phys. 68 (1991) 936-939.
25. D. Hönig and D. Möbius: J. Phys. Chem. 95 (1991) 4590-4592.
26. D. Jacquemain, S. G. Wolf, F. Leveiller, M. Deutsch, K. Kjaer, J. Als-Nielsen, M. Lahav and L. Leiserowitz: Angew. Chem. Int. Ed. Engl. 31 (1992) 130-152.
27. T. Kajiyama: MRS Bulletin June (1995) 32-38.
28. C. M. Knobler and R. C. Desai: Annu. Rev. Phys. Chem. 43 (1992) 207-236.
29. G. A. Overbeck and D. Möbius: J. Phys. Chem. 97 (1993) 7999-8004.
30. S. Rivière, S. Hénon, J. Meunier, D. K. Schwartz, M.-W. Tsao and C. M. Knobler: J. Chem. Phys. 101 (1994) 10045-10051.
31. N. R. Pallas and B. A. Pethica: J. Chem. Soc. Farad. Trans. I 83 (1987) 585-590.
32. P. J. Winch and J. C. Earnshaw: J. Phys.: Condens. Matter 1 (1989) 7187-7205.
33. N. R. Pallas and B. A. Pethica: Lang. 9 (1993) 361-362.
34. B. G. Moore, C. M. Knobler, S. Akamatsu and F. Rondelez: J. Phys. Chem. 94 (1990) 4588-4595.
35. H. M. McConnell, L. K. Tamm and R. M. Weis: Proc. Natl. Acad. Sci. USA 81 (1984) 3249-3253.
36. M. Lösche and H. Möhwald: Rev. Sci. Instrum. 55 (1984) 1968-1972.
37. R. Peters and K. Beck: Proc. Natl. Acad. Sci. USA 80 (1983) 7183-7187.
38. K. A. Suresh, J. Nittmann and F. Rondelez: Europhys. Lett. 6 (1988)
39. B. Moore and C. M. Knobler: J. Chem. Soc., Faraday Trans. 2 82 (1986) 1753-1761.
40. D. Hönig, G. A. Overbeck and D. Möbius: Adv. Mater. 4 (1992) 419-424.
41. D. Hönig and D. Möbius: Thin Solid Films 210/211 (1992) 64-68.

42. V. T. Moy, D. J. Keller, H. E. Gaub and H. M. McConnell: J. Phys. Chem. 90 (1986) 3198-3202.

43. X. Qiu, J. Ruiz-Garcia, K. J. Stine and C. M. Knobler: Phys. Rev. Lett. 67 (1991) 703-706.

44. J. Ruiz-Garcia, X. Qui, M.-W. Tsao, G. Marshall, C. M. Knobler, G. A. Overbeck and D. Mobius: J. Phys. Chem. 97 (1993) 6955-6957.

45. D. K. Schwartz, M.-W. Tsao and C. M. Knobler: J. Chem. Phys. 101 (1994) 8258-8261.

46. D. K. Schwartz and C. M. Knobler: J. Phys. Chem. 97 (1993) 8849-8851.

47. K. Motomura, S. Iwanaga, Y. Hayami, S. Uryu and R. Matuura: J. Coll. Int. Sci. 80 (1981) 32-38.

48. M. Aratono, S. Uryu, Y. Hayami, K. Motomura and R. Matuura: J. Coll. Int. Sci. 93 (1983) 162-168.

49. M. Aratono, S. Uryu, Y. Hayami, K. Motomura and R. Matuura: J. Coll. Int. Sci. 98 (1984) 33-38.

50. K. Lunkenheimer, G. Serrien and P. Joos: J. Coll. Int. Sci. 134 (1990) 407.

51. M. Lin, J.-L. Firpo, P. Mansoura and J. F. Baret: J. Chem. Phys. 71 (1979) 2202-2206.

52. S. Hénon and J. Meunier: J. Chem. Phys. 98 (1993) 9148-9154.

53. S. Lin, K. McKeigue and C. Maldarelli: Langmuir 24 (1991) 159.

54. J. F. Baret: J. Phys. Chem. 72 (1968) 2755-2758.

55. J. F. Baret: J. Colloid and Int. Sci. 30 (1969) 1-12.

56. V. B. Fainerman: Colloid Journal USSR 39 (1977) 91.

57. V. B. Fainerman: Colloid J. USSR 40 (1978) 437.

58. R. P. Borwankar and D. T. Wasan: Chemical Engineering Science 38 (1983) 1637-1649.

59. R. P. Borwankar and D. T. Wasan: Chemical Engineering Science 41 (1986) 199-201.

60. R. v. d. Bogaert and P. Joos: J. Phys. Chem. 83 (1979) 2244-2248.

61. R. v. d. Bogaert and P. Joos: J. Phys. Chem. 84 (1980) 190-194.

62. R. L. Bendure: J. Colloid and Interface Science 35 (1971) 238-248.

63. J. Kloubek: J. Colloid and INterface Sci. 41 (1972) 1-6.

64. K. J. Mysels: Langmuir 2 (1986) 428-432.

65. K. J. Mysels: Colloids and Surfaces 43 (1990) 241-262.

66. P. R. Garrett and R. D. Ward: A reexamination of the measurement of dynamic surface tensions using the maximum bubble pressure method J. Colloid and Interface Science (1989) 475-490.

67. A. Passerone, L. Liggieri, N. Rando, F. Ravera and E. Ricci: J. COlloid and Interface Science 146 (1991) 152.

68. C. A. Macleod and C. J. Radke: J. Colloid and Interface Science 160 (1993) 435-448.

69. X. Zhang, M. Harris and O. Basaran: J Colloid and INterface Science 168 (1994) 47-60.
70. R. Nagarajan and D. T. Wasan: J Colloid and Interface Science 159 (1993) 164-175.
71. L. Liggieri, F. Ravera and A. Passerone: J Colloid and Interface Science 169 (1995) 226-237.
72. E. Tornberg: Journal of Colloid and Interface Science 64 (1978) 391-402.
73. F. W. Pierson and S. Whitaker: Journal of Colloid and Interface Science 54 (1975) 203-218.
74. P. Joos and E. Rillaerts: Journal of Colloid and Interface Science 79 (1981) 96-100.
75. C. Jho and R. Burke: Journal of Colloid and Interface Science 79 (1983) 96.
76. R. Miller, A. Hofmann, R. Hartmann, K. Schano and A. Halbig: Advanced Materials 4 (1992) 370374.
77. R. Miller and K. H. Schano: Colloid and Polymer Science 264 (1986) 277-281.
78. J. v. Hunsel, G. Bleys and P. Joos: Journal of Colloid and Interface Science 114 (1986) 432-441.
79. Y. Rotenberg, L. Boruvka and A. W. Neumann: Journal of Colloid and Interface Science 93 (1983) 169-183.
80. S. Y. Lin, K. McKeigue and C. Maldarelli: AIChE Journal 36 (1990) 1785-1795.
81. Girault, D. Schiffrin and B. D. V. Smith: J. of Colloid and Interface Science 101 (1984) 257-266.
82. P. Cheng, D. Li, L. Boruvka, Y. Rotenberg and A. W. Neumann: Colloids and Surfaces 43 (1990) 151-167.
83. R. Miller and K. Lunkenheimer: Colloid and Polymer Science 264 (1986) 357-361.
84. J. Lucassen and M. v. d. Tempel: Chemical Engineering Science 27 (1972) 1283.
85. J. Lucassen and G. T. Barnes: J. C. S. Faraday I 68 (1972) 2129.
86. K. Lunkenheimer, C. Hartenstein, R. Miller and D. Wantke: 8 (1984)
87. G. Gottier, N. Admunson and R. Flumerfelt: J. of Colloid and Interface Scienc 114 (1986) 106.
88. C. Hsu and R. E. Apfel: Journal of Colloid and Interface Science 107 (1985) 467.
89. C. Chang and E. Franses: Journal of Colloid and Interface Science 164 (1994) 107-113.
90. C. Chang and E. Franses: Chemical Engineering Science 49 (1994) 313-325.
91. R. Miller, R. Sedev, K. H. Schano, C. Ng and A. W. Neumann: Colloids and Surfaces A 69 (1993) 209-216.
92. S. Lin, T. Lu and W. Hwang: Langmuir 11 (1995) 555-562.
93. J. Lucassen and v. d. Tempel: J Colloid and Interface Sci. 41 (1972) 491.
94. C. Sohl, K. Miyano and J. Ketterson: Rev. Sci. Instruments 49 (1978) 1464-1469.
95. C. Stenvot and D. Langevin: Langmuir 4 (1988) 1179-1183.
96. D. Langevin: Light Scattering by Liquid Surfaces, Marcel Dekker, New York, 1991.
97. K. Miyano, B. Abraham, L. Ting and D. T. Wasan: J Colloid and Interface Sci. 92 (1983) 297-302.

98. C. Lemaire and D. Langevin: Colloids and Surfaces (1992)
99. I. Panaiotov, A. Sanfeld, A. Bois and J. F. Baret: J of Colloid and Interface Science 96 (1983) 315-321.
100. G. Loglio, U. Tesei and R. Cini: Rev. Sci. Instr. 59 (1988) 2045.
101. P. D. Keyser and P. Joos: Journal of Colloid and Interface Science 91 (1983) 131-137.
102. P. Joos and G. Bleys: Colloid and Polymer Science 261 (1983) 1038-1042.
103. J. v. Hunsel, D. Vollhardt and P. Joos: Langmuir 5 (1989) 528-531.
104. P. Joos and M. v. Uffelen: Journal of Colloid and Interface Science 155 (1995) 271-282.
105. R. Miller, G. Loglio, U. Tesei and K. H. Schano: Advances in Colloid and Interface Science 37 (1991) 73-96.
106. J. Lucassen and D. Giles: J Chem Soc Faraday Trans I 71 (1975) 217.
107. G. Bleys and P. Joos: J Phys Chem 89 (1985) 1027-1032.
108. R. Defay and J. R. Hommelen: J Colloid Sci. 14 (1959)
109. V. B. Fainerman and S. V. Lylyk: Kolloidn Zh 44 (1982) 598.
110. J. F. Baret, A. G. Bois, L. Casalta, J. Dupin, J. Firpo, J. Gonella, L. Melinon and J. Rodeau: Journal of Colloid and Interface Science 25 (1975) 503.
111. L. Ting, D. T. Wasan, K. Miyano and S. Xu: J Colloid and Interface Science 102 (1984) 248-258.
112. P. Joos and G. Serrien: Journal of Colloid and Interface Science 127 (1989) 97-103.
113. C. Chang and E. Franses: Colloids and Surfaces 69 (1992) 189-201.
114. C. Tsonopoulos, J. Newman and J. M. Prausnitz: Chemical Engineering Science 26 (1971) 817-827.
115. P. Joos, G. Bleys and G. Petre: J. Chim. Phys. 79 (1982) 387.
116. M. D. Levan: Jpornal of Colloid and Interface Science 83 (1981) 11-17.
117. J. Chen and K. Stebe: Journal of Colloid and Interface Science 178 (1996) 144-155.
118. M. Levan and J. Newman: AIChE Journal 22 (1976) 695-701.
119. J. Holbrook and M. L. Levan: Chemical Engineering Communications 20 (1983) 191-207.
120. J. Holbrook and M. L. Levan: Chemical Engineering Communications 20 (1983) 273-290.
121. R. E. Davis and A. Acrivos: Chemical Engineering Science 21 (1966) 681-685.
122. S. S. Sadhal and R. E. Johnson: Journal of Fluid Mechanics 126 (1983) 237.
123. Z. He, C. Maldarelli and Z. Dagan: Journal of Colloid and Interface Science 146 (1991) 442-451.

LAMINAR TRANSPORT OF SOLID PARTICLES SUSPENDED IN LIQUIDS

U. Schaflinger
Technical University, Vienna, Austria

ABSTRACT

Attention is focused on the laminar transport of non-colloidal, non-buoyant particles suspended in a continuous, Newtonian fluid. In the presence of a gravitational force field particles settle on the bottom of the conduit, while hydrodynamic diffusion acts against gravity and causes at least a part of the particles to be resuspended. Resuspension of heavy particles under the action of shear has been observed not only in turbulent flows but also when the Reynolds number is small. Since particle-particle interactions are important, an individual sphere will experience shear-induced diffusion. This phenomenon, which causes an upward flux of particles from regions of high concentrations to low is eventually balanced by the downward flux due to gravity. Several examples of uni-directional resuspension flows have been investigated in both theoretical and experimental levels. Experiments performed in a 2-D Hagen-Poiseuille channel show reasonable agreement with the theory for small flow rates of clear liquid. In certain parameter ranges, however, different flow instabilities were reported: When the flow rates were large the interface between the clear liquid and the suspension became unstable and the measured flow quantities were much different than theoretical predictions. Relatively heavy particles occasionally caused partial blocking of the channel and the clear fluid meandered through the stagnant sediment. Eventually, ripple-type instabilities were found in the case of a very thin layer of particles. A linear stability analysis for a 2-D Hagen-Poiseuille resuspension flow revealed that the interface is almost always unstable. The numerical solutions show that two branches contribute to the convective instability; i.e., long and short waves, which coexist in a certain range of parameters. Also, a large range exists where the flow is absolutely unstable. Finally, the entrance flow of an originally well mixed suspension flowing into a two-dimensional channel and the propagation of a sediment layer can be investigated by applying a theory of kinematic waves.

1. INTRODUCTION

Resuspension of heavy particles is usually associated with turbulent flows and has long been known to play a significant role in many industrial applications and in medicine, such as the transport of slurries and the flow of blood. Gadala-Maria [1], however, appears to have been the first to notice that resuspension can also occur at vanishingly small Reynolds numbers, for which inertial effects are insignificant and the flow is laminar. This phenomenon, as a by-product of investigations regarding the "Rheology of Concentrated Suspension of Non-colloidal Particles" (Acrivos [2]), has been designated "Viscous Resuspension" (Leighton and Acrivos [3]) and has recently attracted extensive attention, e.g., in oil prospecting (Unwin and Hammond [4]). Viscous resuspension is linked to the shear-induced diffusion mechanism for the migration of particles across streamlines in creeping flows. This diffusion mechanism is quite different from Brownian diffusion, which is negligible for particles of the size used throughout this paper. Eckstein et al. [5] determined experimentally the coefficient of lateral self-diffusion in the direction normal to the fluid velocity. Their experimental technique was later improved by Leighton and Acrivos [6] who investigated a wide range of parameters and verified that the coefficient of lateral self-diffusion is proportional to the product of the applied shear rate $\dot{\gamma}$ and the square of the particle radius a^2, the proportionality being a function of the particle volume concentration Φ. Self-diffusion, however, is not the sole mechanism by which a particle can drift across the streamlines of the flow. As a result of particle-particle interactions, a diffusive flux also occurs from regions of high concentrations to low as well as from regions of high shear to low. Leighton and Acrivos [7] showed that the relevant diffusion coefficient again scales with $\dot{\gamma}a^2$.

Consider the uni-directional, laminar flow of a suspension of neutrally buoyant spheres in a tube or channel and suppose that the concentration at the inlet port is homogenous. The concentration profile downstream from the inlet becomes non-uniform due to shear-induced diffusion, i.e., particles migrate from the region of high shear (the wall) to the center, where the shear vanishes. The emerging gradient in concentration, however, opposes this flux until a fully developed flow is established far downstream. When the flow is fully developed, the particle concentration at the centerline of the duct equals the maximum concentration $\Phi = \Phi_0$ (Phillips et al. [8]) and the effective viscosity becomes infinite. The suspension in the central region of the pipe or channel is therefore expected to move as a plug.

If an originally homogenous suspension of heavy spheres flows through a horizontal tube or channel, shear-induced diffusion is eventually balanced by the downward flux due to gravity. In this case, a fully developed flow consists of either two or three distinct regions (Leighton and Acrivos [3], Schaflinger et al. [9]): a clear fluid layer on top, a flowing suspension whose concentration decreases monotonically with height above the bottom wall, and, in some cases, a stagnant sediment layer with maximum particle concentration on the bottom. The thickness of these different regions can be predicted theoretically by modeling the suspension as an effective fluid whose viscosity depends on the particle concentration and then solving simultaneously the appropriate momentum and continuity equations. This shear-induced diffusion phenomenon was first reported by Gadala-Maria [1], who observed that a sediment layer of particles at the bottom of a channel was resuspended by a laminar shear flow of clear fluid above.

Several examples of uni-directional resuspension flows, such as plane Couette flow (Leighton and Acrivos [3]), plane film flow, 2-D Hagen-Poiseuille channel flow (Schaflinger et al. [9]), and quasi-unidirectional resuspension flow (Nir and Acrivos [10], Kapoor and Acrivos [11]) as it occurs in inclined gravity settling, have been investigated in both theoretical and experimental levels. Recently, Acrivos et al. [12] studied the shear-induced resuspension in a narrow gap Couette device where the diffusive flux is normal to the plane of shear. Eventually, Zhang and Acrivos [13] analyzed viscous resuspension of heavy particles in fully developed laminar pipe flows. They extended the model previously used for unidirectional flows to cases in which all three velocity components are non-zero due to the existence of a secondary motion within the cross section of the pipe. The numerical results obtained by Zhang and Acrivos [13] are in excellent agreement with earlier experiments performed by Altobelli et al. [14].

Experiments performed in a Couette device and a 2-D Hagen-Poiseuille channel show reasonable agreement with the theory for relatively small flow-rates of clear liquid and moderate suspension heights (Leighton and Acrivos [3], Schaflinger et al. [15], Acrivos et al. [16]). Within certain ranges of parameters, however, different flow instabilities were reported and the measurements were much different from the theoretical predictions. Larger flow-rates of clear liquid caused interfacial waves that could be very strong. Schaflinger et al. [15] observed wave breaking and clouds of detached particles moving much faster than the suspension layer.

Relatively heavy particles and very concentrated feed suspensions caused an unsteady flow with partial blocking of the channel and the clear fluid meandering through the sediment (Schaflinger et al. [9]). Also, ripple-type instabilities were observed when the layer of particles was very thin (Schaflinger [17], Schaflinger et al. [15]).

Zhang et al. [18] performed a linear stability analysis and found that a 2-D Hagen-Poiseuille resuspension flow is almost always unstable to interfacial waves. They assumed that the particle concentration was uniform throughout the resuspension layer, which is justified by the base-state results even for moderate flow-rates of clear liquid, and they also introduced a locally averaged particle concentration. A subsequent numerical study of this problem (Schaflinger [17]) reveals that the linear stability problem is quite sensitive to the concentration. Since the concentration within the suspension layer is always near the maximum value, even small variations in it have a large effect on the effective viscosity, which in turn drastically influences the stability of the flow. Using the average concentration as additional parameter, Schaflinger [17] revealed the existence of two different convective instabilities: relatively long and short waves, which coexist in a certain range of parameters. Also, a range exists where the flow is absolutely unstable, which means that a convectively unstable resuspension flow can be only observed between a lower and an upper critical value of the Reynolds number. The flow is absolutely unstable above this upper critical value. In some cases, however, there exists an even larger third boundary beyond which the flow is convectively unstable again. Short waves were usually observed in the experiments, but the occasional coexistence of short and long waves was also reported Finally, the linear stability analysis also predicts interfacial waves when the suspension heights are small. These results are in accordance with measurements for ripple-type instabilities as they occur under laminar conditions for a mono-layer of particles (Schaflinger [17], Schaflinger et al. [15]).

Very long waves usually propagate as kinematic waves, where kinematic shocks may also occur (Witham [19], Kluwick [20]). Schaflinger [21] employed the theory of kinematic waves to investigate the propagation of a change of height of the resuspended particle layer. He found that in both fully developed planar Hagen-Poiseuille channel flow and 2-D gravity-driven film flow, a change of suspension-height moves as a combination of kinematic shocks and kinematic waves. The observed motion of a sudden end of a suspension layer in a 2-D Hagen-Poiseuille channel flow agreed well with the theory if the flow rate was small but showed distinct discrepancies from the theoretical predictions for larger flow-rates (i.e., when dynamic waves occurred at the interface).

Eventually, the entrance of an originally well mixed suspension flowing into a two-dimensional channel and the subsequent propagation of the emerging sediment layer can be also investigated by applying the theory of kinematic waves (Schaflinger [23]). It has been shown that particle separation due to gravity is usually the dominating effect at the entrance region. This is because, when a well-mixed but dilute suspension of heavy particles enters the channel, the flux is negligible at first because shear-induced diffusion becomes only important at a later state, i.e., when the flow has already segregated into a pure liquid and a concentrated suspension. Thus, we assume local equilibrium in the spatially growing sediment layer between the flux due to gravity and hydrodynamic diffusion. In such a case, the entrance length is equal to the distance until the resuspended sediment layer has reached its maximum height, i.e., when all particles have separated from the original homogeneous suspension.

2. BASIC CONCEPT

For illustrating the theory we consider a suspension of negatively buoyant spheres of uniform size resuspended in either a 2-D Hagen-Poiseuille channel or in a plane film flow, as depicted in figs. 1a and 1b.

The symbol h_0 denotes the height of the sediment that would be reached if the flow were stopped and the particle layer had reached its maximum volume concentration $\Phi_0 \sim 0.58$. The position of the top of the resuspended layer in the presence of a laminar shear flow with velocity $U(z)$ is denoted by h_t while h_c refers to the height of a possibly remaining sediment layer at the bottom. Q is the volume flux of clear liquid per unit depth, μ is the dynamic viscosity, ρ is the density and ν the kinematic viscosity of the pure liquid. The subscript m refers to the particle-fluid mixture within the resuspended layer where both the viscosity μ_m and the density ρ_m are functions of the particle concentration $\Phi(z)$. Where necessary within the text, subscripts 1 and 2 will distinguish between the clear fluid and the particle properties, respectively. Further, the symbol g refers to the gravitational constant and β denotes the angle of inclination. Finally, the total height of the 2-D duct is given by $2B$ (fig. 1a), while δ is the total thickness of the downward flowing film according to fig. 1b.

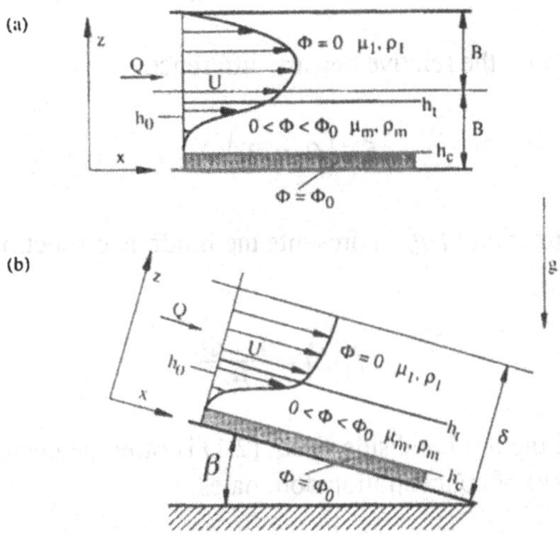

Fig. 1. (a) Schematic of a 2-D Hagen-Poiseuille flow indicating the notation. (b) Plane film flow (from Schaflinger [21]).

For uni-directional fully developed flows the particle flux due to a gradient in concentration and a gradient in shear is balanced by the particle flux due to gravity:

$$\frac{2\Phi a^2 g\,\varepsilon}{9\,\nu_1} f(\Phi) + \dot{\gamma}(z)\,a^2\widehat{D}\,\frac{\mathrm{d}\Phi}{\mathrm{d}z} - \dot{\gamma}(z)\,a^2\widetilde{D}\frac{1}{\tau}\frac{\mathrm{d}\tau}{\mathrm{d}z} = 0, \tag{2.1}$$

with the diffusion coefficients given as (Leighton and Acrivos [7])

$$\widehat{D} = \frac{1}{3}\,\Phi^2\left(1 + \frac{1}{2}\,\mathrm{e}^{8.8\,\Phi}\right) \tag{2.2}$$

and

$$\widetilde{D} = 0.6\,\Phi^2. \tag{2.3}$$

The shear rate is defined as

$$\dot{\gamma} = \tau(z)/\mu_m(z)\,, \tag{2.4}$$

with the effective viscosity written as

$$\mu_r = \frac{\mu_m}{\mu_1} = \left(1 + \frac{1.5\,\Phi}{1 - \Phi/\Phi_0}\right)^2 . \tag{2.5}$$

The symbol ε defines the relative density difference

$$\varepsilon = \left(\rho_2 - \rho_1\right)/\rho_1\,. \tag{2.6}$$

In eq. (2.1) the function $f(\Phi)$ represents the hindrance function which is assumed to be

$$f(\Phi) = \frac{1 - \Phi}{\mu_r}.$$

The third term of the left hand side in eq. (2.1) is often neglected since shear-induced migration due to a gradient of concentration dominates.

The fully-developed flow is governed by the particle concentration Φ_s in the well-mixed entrance region and a modified Shields number which for the 2-D Hagen-Poiseuille flow is

$$\kappa_P = 9\, \nu_1\, Q/16\, B^3\, g\, \varepsilon\,, \tag{2.7}$$

and for the plane film flow is

$$\kappa_F = (\tan\beta)/\varepsilon. \tag{2.8}$$

It is interesting to note that the structure of the fully developed flow is independent of the particle radius a. This fact has been experimentally confirmed by Leighton and Acrivos [3] and Schaflinger et al. [9].

Eq. (2.1) together with the equations of motion and continuity and the appropriate boundary conditions must be solved numerically.

3. APPLICATIONS AND EXPERIMENTAL OBSERVATIONS

In fig. 2 some typical concentration profiles clearly show that the concentration Φ changes rapidly near the suspension-clear fluid interface and hardly at all below it.

Fig. 3a shows velocity profiles for a 2-D Hagen-Poiseuille flow. The theory predicts that for $\kappa_p >> 1$, both the velocity within the clear fluid and the velocity within the resuspended layer are parabolic. In this limit Φ is constant and the shear stress vanishes at the interface. It is interesting to note, however, that already for $\kappa_p = 0.72$ the concentration seems to be constant that is important for the linear stability analysis (Zhang et al. [18], Schaflinger [17]). Fig. 3b depicts three velocity profiles for a plane gravity-driven film flow. In the case $(\tan\beta)/\varepsilon = 0.01746$ the shear stress is not strong enough to resuspend the whole settled sediment and a remaining, stagnant layer appears.

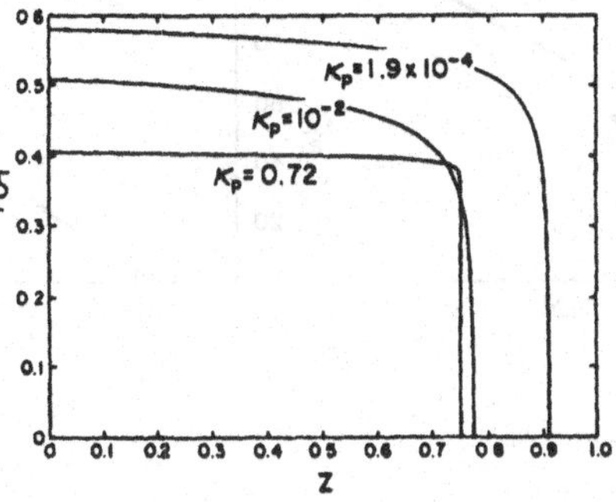

Fig. 2. Particle concentration profiles in a channel flow for $\Phi_s = 0.3$ (from Schaflinger et al. [9])

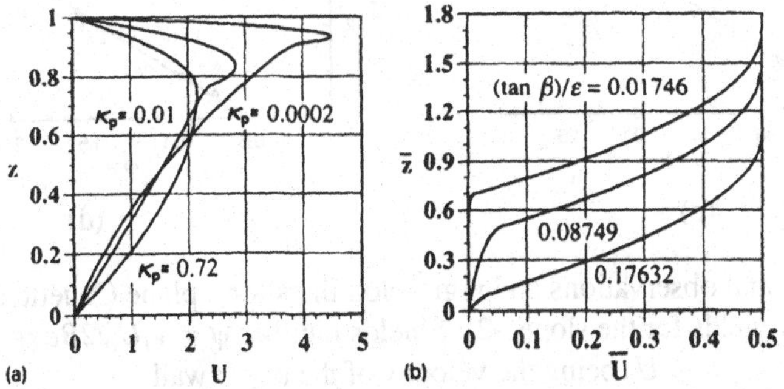

Fig. 3. (a) Velocity profiles for a 2-D Hagen-Poiseuille flow (theoretical predictions, $\kappa_p = 0.0002$, $h_t = 0.912$; $\kappa_p = 0.01$, $h_t = 0.774$; $\kappa_p = 0.72$, $h_t = 0.75$); (b) Velocity profiles for a plane, gravity-driven film flow (theoretical predictions, $\tan\beta/\varepsilon = 0.01746$, $h_t = 0.7$; $\tan\beta/\varepsilon = 0.08749$, $h_t = 0.5$; $\tan\beta/\varepsilon = 0.17632$, $h_t = 0.1$). (from Schaflinger [21])

For a plane Couette flow the experimental results (figs. 4a-4d) show for all tested suspensions an approximately linear relation between the change of height of the resuspended layer $\Delta h / a$ and the Shields number ψ as was expected from theory (Leighton and Acrivos [3]). Here we note that the modified Shields number κ is related to the Shields number ψ via the particle radius a referred to the total height of the channel.

For sedimenting suspensions flowing in a pressure-driven 2D Hagen-Poiseuille channel, an estimate for the entrance length L_ϕ, required for the establishment of a fully developed particle concentration profile, can be obtained readily by balancing the convective flux and that due to sedimentation.

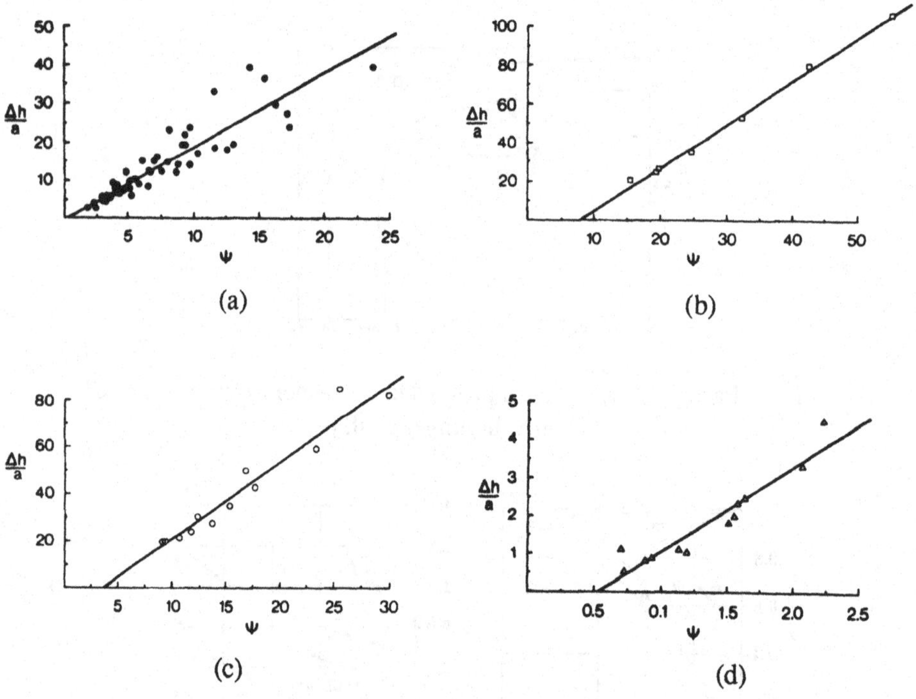

Fig. 4. Experimental observations and regression lines for a plane Couette flow $h_t - h_o = \Delta h$ (theory yields for the slope ~2); Shields number $\psi = v_l U_o / 2Bag\varepsilon$ with U_o being the velocity of the upper wall
(a) 139μm glass spheres; (b) 46μm polystyrene spheres; (c) 43μm glass spheres; (d) Mississippi river sediment. (from Leighton and Acrivos [3])

This is because, when a well-mixed but dilute suspension of heavy particles enters the channel, the flux due to shear-induced diffusion is negligible at first and becomes important only at a later state, i.e., when the flow has already segregated into a pure liquid and a concentrated suspension. Consequently

$$\frac{Q}{B\,L_\phi} \sim \frac{a^2 g\,\varepsilon}{B\,\nu_1} \quad \text{or} \quad L_\phi \sim \kappa \frac{B^3}{a^2}\,. \tag{3.1}$$

This is in contrast to the case of neutrally buoyant particles, where a balance between the convective flux and that due to shear induced diffusion leads to the estimate (Nott and Brady [22])

$$L_\phi \sim \frac{B^3}{a^2}\,. \tag{3.2}$$

In all the experiments, the value of L_ϕ given by equation [3.1] was much smaller than the total length of the channel L, since $\kappa B^3/a^2$ never exceeded approximately 5 cm.

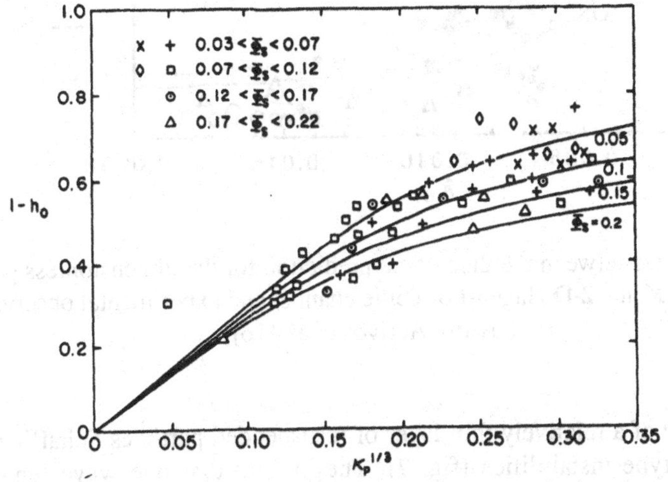

Fig. 5. Settled height h_0 vs. κ_p for different values of Φ_s and comparison with experiments for 2-D Hagen-Poiseuille channel flow (+, □, O, Δ, polystyrene beads; x, ◊, glass beads). (from Schaflinger et al. [9])

In fig. 5 observations of the settled height h_0 in a 2-D Hagen-Poiseuille channel and corresponding theoretical predictions are plotted versus $\kappa_p{}^{1/3}$ (Schaflinger et al. [9]). For larger values of κ_p the measurements show a relatively large degree of scatter due to the existence of interfacial waves. The influence of instabilities on a resuspension flow in a 2-D

Hagen-Poiseuille is also clearly visible in fig. 6 (Acrivos et al. [16]). Here, the measured and theoretically predicted dimensionless pressure drop coefficient $K/12$ is plotted as a function of κ_p with Φ_s being a parameter.

In all cases interfacial flow instabilities were observed for $\kappa_p > 0.005$. Since the interfacial waves are believed to lead to a higher degree of resuspension, the theory usually overestimates the pressure drop. These observations provided the motivation for a linear stability analysis that will be described in more detail in the next section.

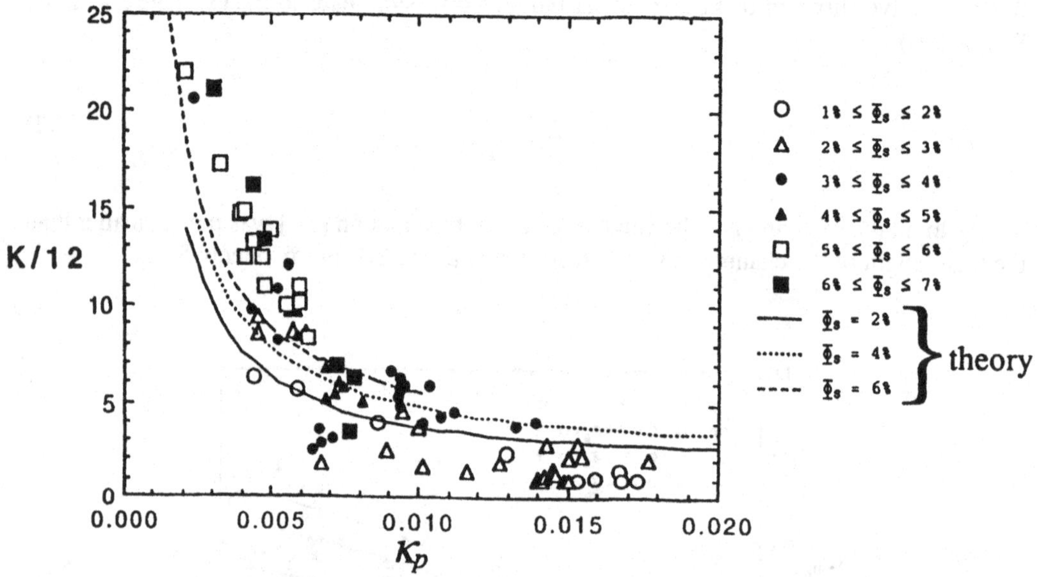

Fig. 6. Comparison between the theoretical prediction for the dimensionless pressure drop coefficient K in a 2-D Hagen-Poiseuille channel and experimental observations. (from Acrivos et al. [16])

In the case of a relatively thin layer of resuspended particles Schaflinger et al. [15] observed ripple type instabilities (fig. 7). They found that one wave length is always approximately 10 particle diameters, a value that was found to be independent of the flow rate of clear liquid. Furthermore, for relatively heavy particles ($\varepsilon \sim 1$) Schaflinger et al. (1990) observed an unsteady flow causing a partial blocking of the channel and the clear liquid meandering through the sediment as sketched in fig. 8.

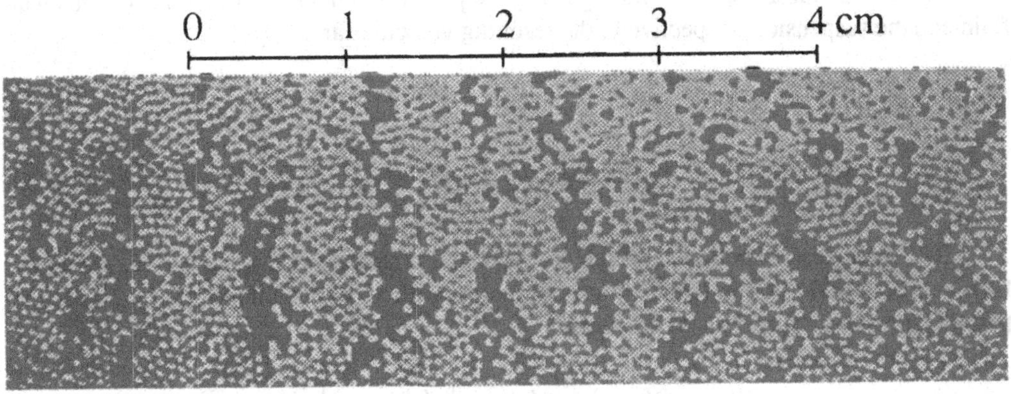

Fig. 7. Top view of a ripple type instability. (from Schaflinger [17])

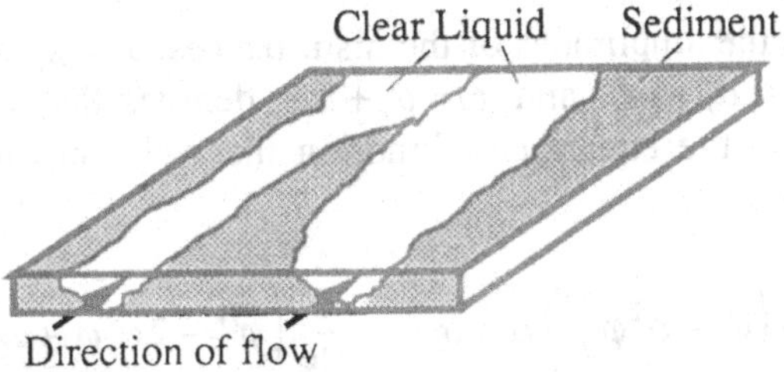

Fig. 8. Sketched observation of a clear liquid meandering through the sediment.

4. HYDRODYNAMIC STABILITY OF RESUSPENSION FLOWS

A linear stability analysis of a shear-induced resuspension flow in a 2-D Hagen-Poiseuille channel should provide at least qualitative information of the conditions at which the flow becomes unstable to interfacial waves. Both the investigation by Zhang et al. [18] and the subsequent studies by Schaflinger [17] are based on a two-fluid model in which the non-uniformity of the particle concentration within the suspension layer is ignored. While Zhang et al. [18] defined a constant concentration as the local mean of the basic concentration profile (fig. 2), Schaflinger [17] kept the mean concentration as an additional parameter since any averaged mean concentration is physically questionable and since the particle concentration within the resuspended layer has a major influence on interfacial waves. In a resuspended particle layer the concentration is typically about 50% and even a small variation of Φ leads to a drastic change of the effective viscosity.

For 2-D disturbances, Ψ_1 and Ψ_2 being the perturbation stream functions of the clear fluid and the suspension, respectively, the resulting velocities are:

$$u_1 = U_1 + \Psi_{1,y};\ v_1 = -\Psi_{1,x}, \tag{4.1}$$

$$u_2 = U_2 + \Psi_{2,y};\ v_2 = -\Psi_{2,x}. \tag{4.2}$$

The linearized equations and the appropriate boundary conditions for the disturbed flow can be represented in terms of normal modes with an exponential time factor for the perturbation stream functions of the form:

$$\Psi_1(x,y,t) = \varphi(y)\exp\left[\mathrm{i}\left(\alpha x - \omega t\right)\right], \tag{4.3}$$

$$\Psi_2(x,y,t) = \psi(y)\exp\left[\mathrm{i}\left(\alpha x - \omega t\right)\right], \tag{4.4}$$

where $\varphi(y)$ and $\psi(y)$ are the amplitudes of the disturbances. The symbol α denotes the complex wave number $\alpha = \alpha_r + \mathrm{i}\alpha_i$ and $\omega = \omega_r + \mathrm{i}\omega_i$ denotes the complex frequency. Eventually we obtain from the equations of motion the well known Orr-Sommerfeld equations for the clear fluid

$$\left(\alpha U_1 - \omega\right)\left(\varphi'' - \alpha^2\varphi\right) - \alpha U_1''\varphi = -\frac{\mathrm{i}}{\mathrm{Re}}\left(\varphi^{IV} - 2\alpha^2\varphi'' + \alpha^4\varphi\right), \tag{4.5}$$

and the suspension

$$\left(\alpha U_2 - \omega\right)\left(\psi'' - \alpha^2\psi\right) - \alpha U_2''\psi = -\frac{\mathrm{i}}{\left(1 + \Phi\varepsilon\right)\mathrm{Re}}\mu\left(\psi^{IV} - 2\alpha^2\psi'' + \alpha^4\psi\right), \tag{4.6}$$

respectively. The boundary conditions for no slip and no penetration at the walls and continuity of velocity and stress at the interface are given in linearized form:

$$\varphi = \varphi' = 0 \quad \text{at} \quad y = 1, \tag{4.7a}$$

$$\psi = \psi' = 0 \quad \text{at} \quad y = 0, \tag{4.7b}$$

$$\varphi - \psi = 0 \quad \text{at} \quad y = h_t. \tag{4.7c}$$

$$\varphi' - \psi' = \alpha\left(U_1' - U_2'\right)\frac{\varphi}{\alpha U_1 - \omega} \quad \text{at} \quad y = h_t, \tag{4.7d}$$

$$\varphi'' - \mu \psi'' = \alpha^2 \varphi (\mu - 1) + \frac{\alpha \varphi}{\alpha U_1 - \omega} \left(U_1'' - \mu U_2'' \right) \quad \text{at} \quad y = h_t , \tag{4.7e}$$

$$i \Phi \varepsilon \alpha^2 \frac{\text{Ga}}{\text{Re}} \frac{\varphi}{\alpha U_1 - \omega} = \varphi''' + \mu \left(3 \alpha^2 \psi' - \psi''' \right) - 3 \alpha^2 \varphi' +$$

$$i \,\text{Re} \left\{ \left(\alpha U_1 - \omega \right) \left[\left(1 + \Phi \varepsilon \right) \psi' - \varphi' \right] - \alpha \left[\left(1 + \Phi \varepsilon \right) U_2' \psi - U_1' \varphi \right] \right\} \tag{4.7f}$$

$$\text{at} \quad y = h_t .$$

Here $\text{Re} = Q / \nu_1$ is the Reynolds number and $\text{Ga} = 8B^3 g / \nu_1^2$ is the Galilei number expressing the effect of gravity g on the stability problem. The symbol ν_1 denotes the kinematic viscosity of the clear liquid. The Reynolds number Re is related to κ_p via $\text{Re} = 2\kappa_p \varepsilon \,\text{Ga}/9$.

The eqs. (4.5 and 4.6) and (4.7a – 4.7f) incorporate a well-known eigenvalue problem that can be solved by any standard method.

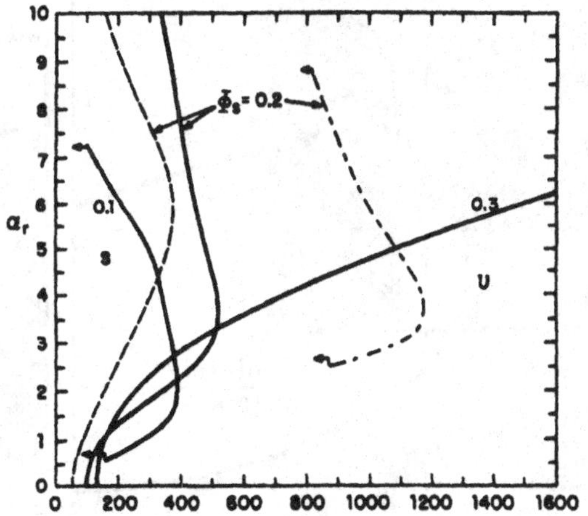

Fig. 9. Neutral stability diagram for $\varepsilon = 10^{-2}$ (solid curves); $\varepsilon = 10^{-3}$ (dashed curve); $\varepsilon = 10^{-1}$ (dash-dot-dashed curve). Stable regions are denoted by S, unstable regions by U. Arrows represent the values of Re below which a sediment layer forms in the base state. (from Zhang et al. [16])

Fig. 9 depicts the topology of the neutral stability curves in the Re – α_r plane for different values of Φ_s. The arrows to the left indicate that a further decrease in Re will lead to the formation of a sediment layer at the bottom of the 2-D duct. The neutral curves consist of a lower and an upper branch (for $\Phi_s = 0.3$, the upper branch is located in the region of large wavenumber and is not shown in fig. 9); the turning point occurs at a value of Re that increases with an increase in the parameters Φ_s and ε. Thus, the wavelengths of the unstable waves are within two distinct ranges, long and short, respectively.

In fig. 10 the bounds for the beginning of instability are shown for different values of Φ (Schaflinger [17]). Again, the upper branch belongs to a long wave instability while the lower branch is associated with short waves. Since the two branches can be continued after intersecting, there is a large region where both types of waves coexist. Fig. 10 also depicts complete mapping of the onset of absolute instability for an otherwise convectively unstable resuspension flow. The region of absolute instability has been plotted for three different values of the mean concentration Φ.

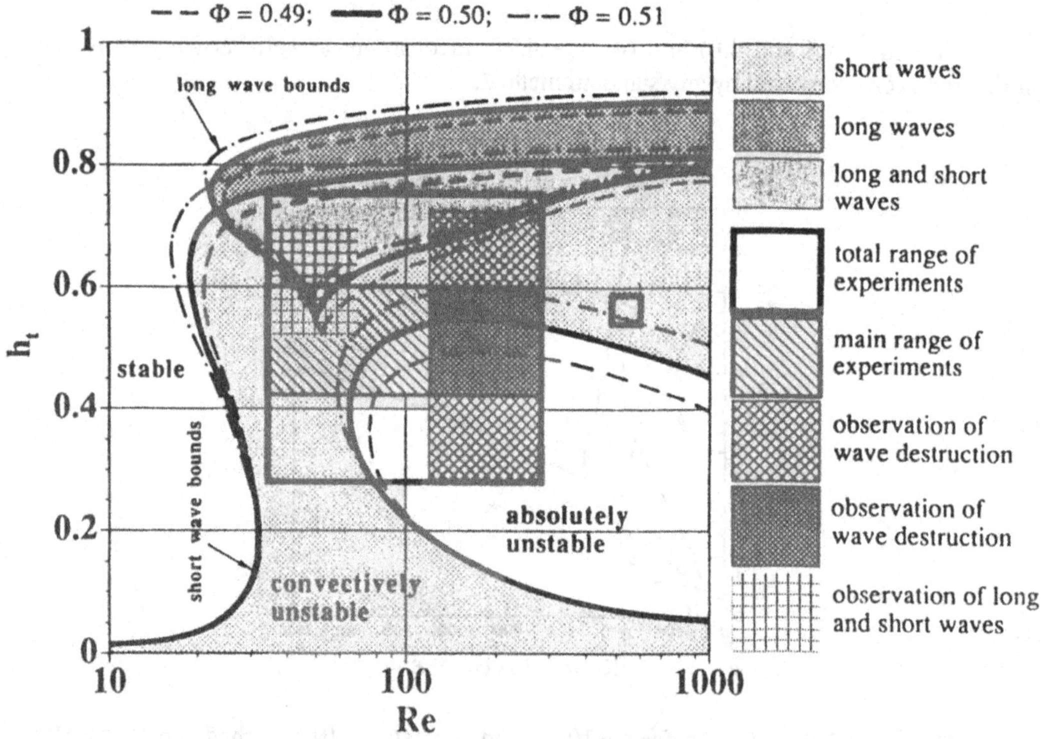

Fig. 10. Stability chart showing the onset of instability and the range of absolute instability. (from Schaflinger [15])

As reported by Schaflinger [17] relatively short and relatively long interfacial instabilities were observed to coexist in the range of small Reynolds numbers. An increase in Re caused the long waves to disappear and eventually very strongly amplified short waves remained which produced wave breaking and clouds of detached particles (fig. 11). Within the latter range of parameters (Acrivos et al., [15]) the measured pressure drop was usually smaller than the theoretical prediction (fig. 6). Hence, we infer that interfacial waves generate a higher degree of resuspension than theoretically predicted by the base flow results. The experiments (Acrivos et al. [16], Schaflinger et al. [15]) also showed that a further increase in Re brought about a transition to smoother interfacial waves.

Even though the mechanism of ripple type instabilities, as they were observed for a very thin particle layer, is not yet fully understood it is interesting to note that the linear stability analysis yielded interfacial waves with wave length that corresponded very well with the measured data (Schaflinger [17]).

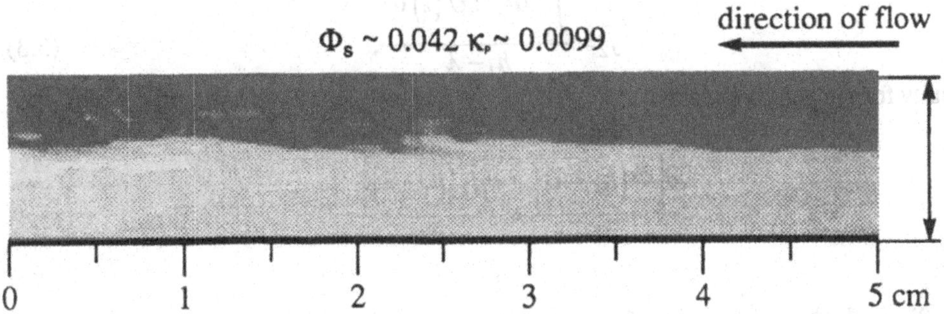

Fig. 11. Photograph of an unstable resuspension flow in a 2-D Hagen-Poiseuille channel. (from Schaflinger et al. [15])

5. PROPAGATION OF KINEMATIC WAVES AND COMPARISON WITH EXPERIMENTS

Very long waves usually propagate as kinematic waves (Witham [19], Kluwick [20]). By neglecting dynamic effects and by means of the theory of kinematic waves Schaflinger [21] investigated the propagation of a change of height,

$$h_0 = h_0(x, t), \tag{5.1}$$

e.g., the beginning (onset) or the end of a resuspended particle layer. He employed the earlier mentioned model for the base-state flow (section 2) and found that both a sudden onset and a sudden end of a resuspended layer in either a fully developed Hagen-Poiseuille channel flow or a plane gravity-driven film flow move as a combination of kinematic shocks and kinematic waves.

In terms of average quantities for the volume concentration

$$\bar{\Phi}=\frac{\int_{\lambda}^{h_t}\Phi(z)\,U(z)\,\mathrm{d}z}{\int_{\lambda}^{h_t}U(z)\,\mathrm{d}z} \tag{5.2}$$

and the particle flux within the resuspended layer

$$\bar{j}_2=\frac{\int_{\lambda}^{h_t}\Phi(z)\,U(z)\,\mathrm{d}z}{h_t-\lambda}, \tag{5.3}$$

continuity for the particles yields

$$\frac{\partial\left[\bar{\Phi}\left(h_t-\lambda\right)\right]}{\partial t}+\frac{\partial\left[\bar{j}_2\left(h_t-\lambda\right)\right]}{\partial x}=0. \tag{5.4}$$

Since $\bar{\Phi}\ (h_t-\lambda)$ and $\bar{j}_2\ (h_t-\lambda)$ are only functions of h_0, eq. (5.4) can be rewritten as a kinematic wave equation for the settled height $h_0(x, t)$:

$$\frac{\partial h_0}{\partial t}+c\,\frac{\partial h_0}{\partial x}=0, \tag{5.5}$$

where the symbol c indicates the kinematic wave velocity

$$c=\frac{\mathrm{d}\left[\bar{j}_2\left(h_t-\lambda\right)\right]}{\mathrm{d}\left[\bar{\Phi}\left(h_t-\lambda\right)\right]}. \tag{5.6}$$

Discontinuities in the sediment height propagate as kinematic shock waves. Such discontinuities can occur at a change of height of the sediment layer. Since inertial effects are neglected, the conservation of particles, in a frame of reference in which the discontinuity is at rest, yields the propagation velocity of the shock

$$W = \frac{\bar{j}_2\left(h_t - \lambda\right) - \hat{j}_2\left(\hat{h}_t - \hat{\lambda}\right)}{\bar{\Phi}\left(h_t - \lambda\right) - \hat{\Phi}\left(\hat{h}_t - \hat{\lambda}\right)}. \tag{5.7}$$

Quantities with a circumflex refer to the state immediately behind the discontinuity while quantities without a circumflex denote variables in front of the shock. In the equations before, the symbol $\lambda = h_c$ if $h_c \geq 0$, while $\lambda = 0$ if the whole sediment has resuspended.

			experimental results			theoretical results		
No.	κ_p x10^{-3}	Re	h_0	h_t	W	h_t	W	comments
1	1.64	38	0.65	0.70	0.009	0.69	0.009	
2	2.44	57	0.26	0.28	0.004	0.27	0.004	
3	2.71	63	0.65	0.70	0.350	0.73	0.365	
4	2.82	65	0.28	0.30	0.005	0.29	0.005	
5	3.05	71	0.62	0.65	0.240	0.69	0.247	
6	3.36	78	0.67	0.75	0.740	0.76	0.730	
7	3.81	88	0.34	0.37	0.012	0.36	0.011	
8	4.21	98	0.21	0.22	0.006	0.22	0.006	slightly wavy
9	4.51	104	0.52	0.60	0.120	0.67	0.077	slightly wavy
10	4.69	109	0.32	0.36	0.026	0.34	0.012	wavy
11	5.26	122	0.37	0.42	0.021	0.40	0.020	
12	5.31	123	0.21	0.30	0.018	0.23	0.007	wavy
13	6.12	142	0.25	0.31	0.020	0.27	0.011	wavy
14	7.49	173	0.24	0.32	0.079	0.27	0.013	very wavy
15	7.56	175	0.48	0.60	0.180	0.56	0.185	wavy
16	7.60	176	0.35	0.42	0.040	0.39	0.029	very wavy
17	9.53	220	0.32	0.45	0.087	0.37	0.040	very wavy
18	10.03	239	0.55	0.60	0.660	0.67	0.676	very wavy
19	11.96	277	0.35	0.40	0.373	0.42	0.099	very wavy
20	12.21	283	0.25	0.26	0.134	0.30	0.038	very wavy
21	12.45	288	0.25	0.27	0.118	0.30	0.041	very wavy
22	14.05	325	0.30	0.36	0.114	0.36	0.077	wavy

Table 1. Comparison of theory and experiment for the propagation of a discontinuity in a 2-D Hagen-Poiseuille flow. (from Schaflinger [21])

According to eq. (5.6) and (5.7) the kinematic wave velocity c and the kinematic shock velocity W, respectively, can be immediately derived from the particle flux $\bar{j}_2$ $(h_t-\lambda)$ vs. concentration $\bar{\Phi}$ $(h_t-\lambda)$ (figs. 12a and 12b) (Schaflinger [21]).

For completeness figs. 13a and 13b show the resuspended height h_t and the stagnant height h_c as a function of h_0 for the cases as depicted in figs. 12a and 12b.

We notice from figs. 12b that for the case of a plane film flow the average particle flux reaches a maximum, constant value as soon as a stagnant layer $h_c \geq 0$ appears. This fact also limits the average concentration because a stagnant layer only leads to a passive displacement of the whole flow. If we discuss a case $\kappa_F = 0.17632$ and an initial settled height $h_0 = 6.0$ and assume a sudden onset and a sudden end of the sediment layer we find from fig. 12b the appropriate particle flux.

Immediately before and after the sediment layer the flux becomes zero, and since the diagram (fig. 12b) shows an inflection point, we defer that at the leading edge and the trailing edge of the sediment layer a kinematic wave and a kinematic shock appear. We also notice that the discontinuity at the onset travels faster than the kinematic wave at the end. Hence, after a certain time, the settled height becomes everywhere smaller than the original one. For the example of an initial layer length of $x = 0.4$ and a sediment height $h_0 = 6.0$ this dimensionless time becomes $t = 1.14$ (fig. 14) (Schaflinger [21]).

Eventually, the leading edge kinematic shock wave totally disappears when it approaches the end of the sediment layer. Finally, the sediment stretches out to a very thin layer that begins at $x = 0$ as a kinematic wave and ends abruptly with a discontinuity (fig. 14). For any initial sediment distribution $h_0 = (x,0)$ this wave form is the final one.

Table 1 shows results for the propagation velocity of a kinematic shock, W, and the height of the resuspended layer, h_t. The measurements for small values of κ_p (runs no. 1-7 in table 1) and for moderate values of h_0, are in relatively good agreement with the theoretical predictions. In some cases (runs no. 9, 10, 12-14,16, 17 and 19-21 in table 1), however, the measured kinematic shock velocity is noticeably larger than the theoretical result. For $\kappa_p > 0.005$ Schaflinger [21] observed interfacial waves that were rather strong during several experimental runs. Those waves, together with wave-breaking and detached particles, as described in chapter 4, led to a layer of particles that moved much faster than the resuspended sediment below. Schaflinger [21] also noticed for larger κ_p that the originally straight discontinuity quickly distorted and moved in jerks rather than continuously.

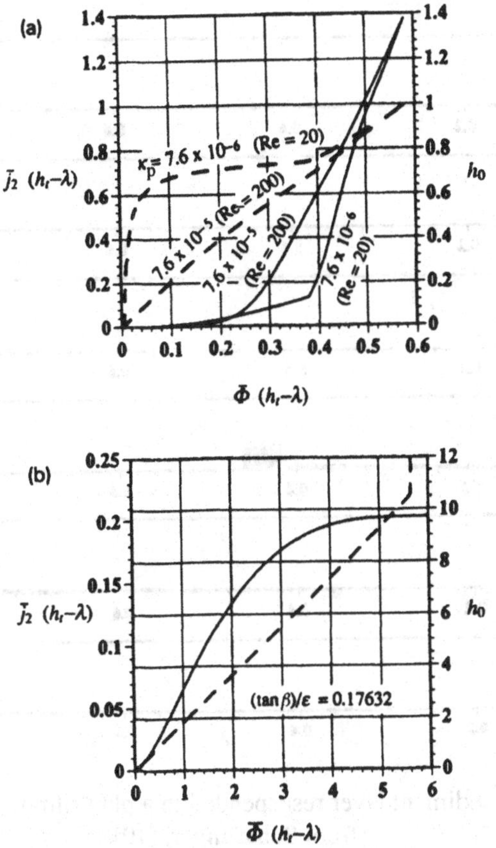

Fig. 12. Particle flux (full lines) and settled height (dashed lines) for: (a) a 2-D Hagen-Poiseuille flow; (b) a plane, gravity driven film flow. (from Schaflinger [21])

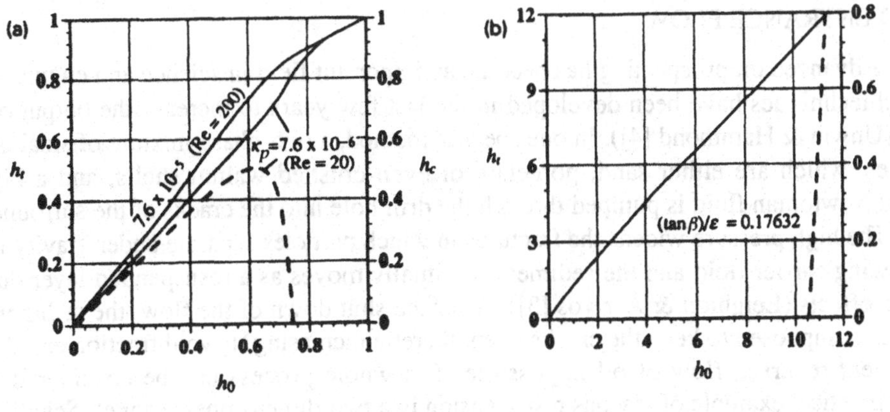

Fig. 13. Resuspended height (full lines) and stagnant height (dashed lines) for: (a) a 2-D Hagen-Poiseuille flow; (b) a plane, gravity driven film flow. (from Schaflinger [21])

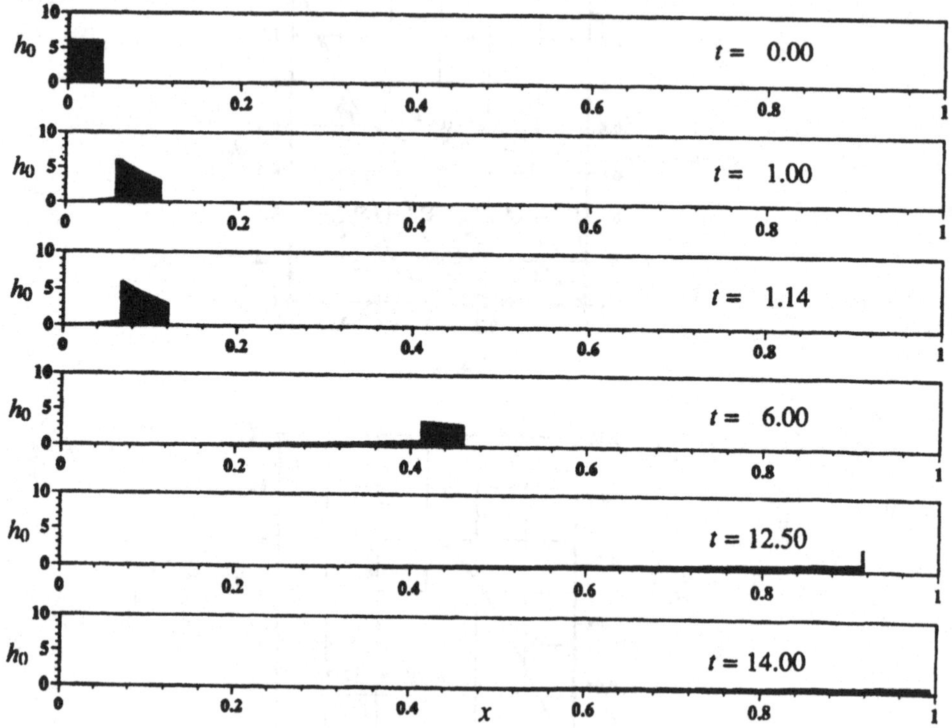

Fig. 14. Motion of a sediment layer resuspended in a plan film flow ($\tan \alpha/\varepsilon = 0.17632$). (from Schaflinger, [19])

6. 2-D ENTRANCE FLOW

Advanced oil prospecting has become an important field in science and engineering. Several techniques have been developed in the last few years to increase the output of oil fields (Unwin & Hammond [4]). In one specific method, a two-phase mixture of small solid particles, which are either sand, porcelain or even crushed walnut shells, and a highly viscous Newtonian fluid is pumped through the drill hole into the cracks of the surrounding rocks. The high pressure widens the fractures in which particles separate under gravity from the flowing carrier fluid and the sediment eventually moves as a resuspended layer due to viscous effects (Leighton & Acrivos [3]). After the shut down of the flow, the sediment of particles is supposed to keep the cracks open, thereby increasing its void fraction and thus a subsequent reversed flow of oil is possible. The whole process can be considered as a typical practical example of viscous resuspension in a two-dimensional channel (Schaflinger et al. [9]) since all Reynolds numbers associated with this flow are small. In order to control the process, it is very important to estimate the entrance region until all particles have separated from the carrier fluid and to determine the distance the sediment layer has traveled within the cracks (Schaflinger, [19]).

As stated earlier, settling of particles is much faster than hydrodynamic diffusion. Thus we assume that we have in the entrance region a sediment resuspended by shear, above a shrinking domain of well-mixed suspension where settling occurs and on top a flow of clear liquid. In terms of averaged quantities continuity for the resuspended sediment requires

$$\frac{\partial(h_t\bar{\Phi})}{\partial t}+\bar{j}_x\frac{\partial h_t}{\partial x}+h_t\frac{\partial \bar{j}_x}{\partial x}=-\bar{j}_{2y}, \qquad (6.1)$$

where $\bar{j}_x$ is the total flux in x-direction and $\bar{j}_{2y}$ denotes the particle flux due to gravity. The flux $\bar{j}_{2y}$ is constant and given by Stokes' law multiplied by a hindrance function and the particle concentration Φ_s.

Since both $\bar{\Phi}$ and $\bar{j}_x$ are only functions of h_t eq. (6.1) can be rewritten as

$$\left(\bar{\Phi}+h_t\frac{d\bar{\Phi}}{dh_t}\right)\frac{\partial h_t}{\partial t}+\left(\bar{j}_x+h_t\frac{d\bar{j}_x}{dh_t}\right)\frac{\partial h_t}{\partial x}=-\bar{j}_{2y}. \qquad (6.2)$$

Eq. (6.2) is again a kinematic wave equation where all derivatives with respect to h_t are known from previous calculations. If we are only interested in steady state solutions eq. (6.2) can be easily numerically integrated and yields the interface between resuspended sediment and still well mixed suspension.

As a boundary condition we assume $h_t = 0$ at $x = 0$. The interface between well mixed suspension and clear liquid h_s simply follows from integrating

$$\frac{dh_s}{dx}=\frac{\bar{j}_{2y}}{\bar{j}_x(h_s)}. \qquad (6.3)$$

Fig. 15 shows the entrance flow for a well-mixed particle concentration $\Phi_s = 0.05$ and a Reynolds number Re = 100. The entrance region is terminated by the intersection of the two interfaces.

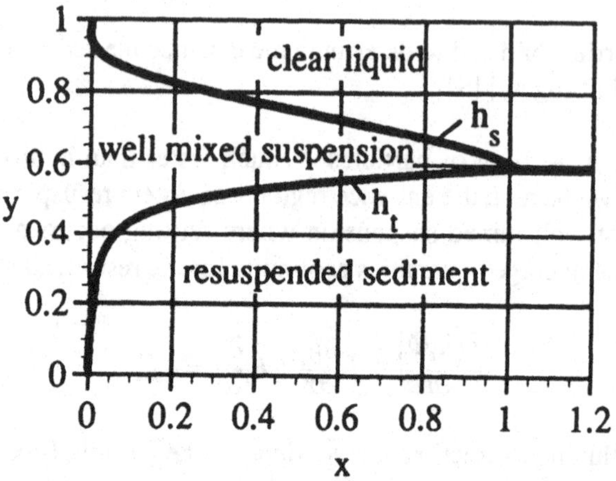

Fig. 15. Entrance flow of a well mixed suspension entering a 2-D Hagen-Poiseuille channel Re = 100, Φ_s = 0.05.

7. SUMMARY

The lecture on laminar transport of solid particles suspended in liquids (viscous resuspension) was meant to highlight the state of art regarding physical understanding gained by theory and experiment. Although the basic mechanisms of shear-induced laminar resuspension might be well understood (Acrivos [2]), no theoretical calculation of the hydrodynamic diffusion coefficients from basic principles has been made yet, mainly owing to the difficulty of analyzing the hydrodynamic interaction of more than two particles. Recently, Acrivos et al. [24] investigated on theoretical bases the coefficient of shear-induced diffusion in the direction of the fluid velocity in a simple shear flow for low particle volume concentration. Thus, only a few uni-directional cases of viscous resuspension, such as plane Couette flow (Leighton and Acrivos [3]), 2-D Hagen-Poiseuille flow, plane gravity-driven film flow (Schaflinger et al.[9]) and quasi-unidirectional resuspension flow (Nir and Acrivos [10], Kapoor and Acrivos [11]) as it occurs in inclined gravity settling, have been studied in detail. Zhang and Acrivos [13] extended the previously used model to theoretically investigate viscous resuspension in fully developed laminar flows in a horizontal pipe. All investigations were based on a balance between shear-induced diffusion due to a gradient in concentration and the particle flux due to gravity.

Experiments (Altobelli [14], Schaflinger et al. [15], Acrivos et al. [16]) agreed well with the theoretical predictions (Leighton and Acrivos [3], Schaflinger et al. [9], Zhang and Acrivos [13]) but it was observed that in 2D Hagen-Poiseuille channel flows larger flow-rates caused different flow instabilities. Schaflinger et al. [15] observed waves at the interface between suspension and clear liquid, ripple type instabilities (Schaflinger et al. [15], Schaflinger [17]) and the clear liquid meandering through the sediment of a partially blocked channel (Schaflinger et al.[9]). A linear stability analysis performed by Zhang et al. [18] and Schaflinger [17] brought to light that a resuspension flow in a 2-D Hagen-Poiseuille channel is almost always unstable to interfacial waves. More clearly, long and short waves occur which can coexist within a certain range of parameters. Also, a region exists where a resuspension flow is absolutely unstable (Schaflinger [17]). In the latter range Schaflinger et al. [15] observed wave-breaking and clouds of detached particles flowing above the original interface. Also, measurements for the pressure drop indicate that resuspension is stronger than predicted if the flow is unstable to interfacial waves.

Since long waves usually travel as kinematic waves (Witham [19], Kluwick [20]), Schaflinger [21] investigated the propagation of a change of height of the resuspended sediment by employing the theory of kinematic waves. He found that in both 2-D Hagen-Poiseuille channel flow and planar gravity-driven film flow changes of the height of sediment travel as a combination of kinematic shocks and kinematic waves. The kinematic shock velocity and the kinematic wave velocity could be easily obtained by means of averaged flow quantities and an appropriate flux diagram. Furthermore, Schaflinger [23] also studied the entrance flow of an originally well mixed suspension flowing into a two-dimensional channel and the subsequent propagation of the emerging sediment layer by applying the theory of kinematic waves. The resulting entrance length is of great importance for advanced oil-prospecting (Unwin and Hammond [4]).

For the future it would be desirable to extend the basic theory to more general flow situations. Also, a nonlinear stability analysis should provide information if interfacial waves indeed cause a higher degree of resuspension, as it is inferred from experimental observations.

REFERENCES

1. Gadala-Maria, F.: The rheology of concentrated suspensions, *Ph.D. Thesis*, Stanford University, CA 1979.

2. Acrivos, A.: The Rheology of Concentrated Suspensions of Non-colloidal Particles, Particulate Two-Phase Flow (Ed. M. Roco), Butterworth-Heinemann, Boston 1993, 169-189.

3. Leighton, D. and A. Acrivos: Viscous resuspension, Chem. Engng. Sci., **41** (1986), 1377-1384.

4. Unwin, A.T. and P.S. Hammond: A simple phenomenological model for shear driven particle migration in concentrated viscoelastic suspension, IUTAM Symposium on Liquid-Particle Interactions in Suspension Flow, Grenoble, France 1994.

5. Eckstein, E.C., D.G. Bailey and A.H. Shapiro: Self-Diffusion of particles in shear flow of a suspension, J. Fluid Mech., **79** (1977), 191-208.

6. Leighton, D. and A. Acrivos: Measurement of shear-induced self-diffusion in concentrated suspensions of spheres, J. Fluid Mech., **177** (1987), 109-131.

7. Leighton, D. and A. Acrivos: The shear-induced migration of particles in concentrated suspensions, J. Fluid Mech., **181** (1987), 415-439.

8. Phillips, R.J., R.C. Armstrong and R.A. Brown: A constitutive equation for concentrated suspensions that accounts for shear-induced particle migration, Phys. Fluids A, **4** (1992), 30-40.

9. Schaflinger, U., A. Acrivos,. and K. Zhang: Viscous resuspension of a sediment within a laminar and stratified flow, Int. J. Multiphase Flow, **16** (1990), 567-578.

10. Nir, A. and A. Acrivos: Sedimentation and sediment flow on inclined surfaces, J. Fluid Mech., **212** (1990), 139-153.

11. Kapoor, B. and A. Acrivos: Sedimentation and sediment flow in settling tanks with inclined wall, J. Fluid Mech., **290** (1995), 30-66.

12. Acrivos, A., R. Mauri and X. Fan: Shear-induced resuspension in a Couette device, Int. J. Multiphase Flow, **19** (1993), 797-802.

13. Zhang, K. and A. Acrivos: Viscous resuspension in fully developed laminar pipe flows, Int. J. Multiphase Flow, **20** (1994), 579-591.

14 Altobelli, S.A., R.C. Givler and E. Fukushima: Velocity and concentration measurements of suspensions by nuclear magnetic resonance imaging, J. Rheol., **35** (1991), 721-734.

15. Schaflinger, U., A. Acrivos and H. Stibi: An experimental study of viscous resuspension in a pressure-driven plane channel flow, Int. J. Multiphase Flow, **21** (1995), 693-704.

16 Acrivos, A., U. Schaflinger, J. Smart and H. Stibi: Measurements in a 2-D Hagen Poiseuille resuspension flow, *Internal report*. Benjamin Levich Institute of PCH, City College of the City University of New York 1993.

17. Schaflinger, U.: Interfacial instabilities in a stratified of two superposed flow, Fluid Dynamics Research, **13** (1994), 299-316.

18 Zhang, K., A. Acrivos and U. Schaflinger: Stability in a two-dimensional Hagen-Poiseuille resuspension flow, Int. J. Multiphase Flow, **18** (1992), 51-63.

19. Witham, G.B.: Linear and Nonlinear Waves, Wiley, New York 1974.

20. Kluwick, A.: Kinematische Wellen, Acta Mechanica, **26** (1977), 15-46.

21. Schaflinger, U.: Transport of a sediment layer due to a laminar, stratified flow, Fluid Dynamics Research, **12** (1993), 95-105.

22. Nott, P.R. and J.F. Brady: Pressure-driven flow of suspensions: simulation and theory, J. Fluid Mech., **275** (1994), 157-199.

23. Schaflinger, U: Motion of a sediment due to a laminar, stratified flow, IUTAM Symposium on Hydrodynamic Diffusion of Suspended Particles, Estes Park, Colorado, USA 1995.

24. Acrivos, A., G.K. Batchelor, E.J. Hinch, D.L. Koch and R. Mauri: Longitudinal shear-induced diffusion of spheres in a dilute suspension, J. Fluid Mech., **240** (1992), 651-657.

RECENT DEVELOPMENTS IN THE ANALYSIS OF GRAVITY AND CENTRIFUGAL SEPARATION OF NON-COLLOIDAL SUSPENSIONS AND UNFOLDING CHALLENGES IN THE CLASSIC MECHANICS OF FLUIDS

M. Ungarish
Israel Institute of Technology, Haifa, Israel

ABSTRACT

This lecture considers the methodology used in the analysis of suspensions that separate under the action of a gravity or a centrifugal field. First some typical results are reviewed, next some new challenging questions (and some answers) that appear in this context in the realm of the classic single-phase fluid theory are discussed. The objective is to increase the audience's apprehansion of these problems and to motivate enhancement of the pertinent research and applications.

1. INTRODUCTION

The motion performed by a "two-phase" mixture of particles (droplets, bubbles) and a suspending fluid, in the presence of a force field, remains very much on the frontier of fluid flow research. Its effective modeling, simulation and interpretation is a difficult task, whose results are bound to be highly beneficial in separation and mixing technologies. For instance, gravity settlers are used in clarification of waste water and processing of mineral particles suspended in mining fluids, and centrifugal machines are essential in separation of milk-cream, purification of contaminated lubrication oils, extraction of viral material in vaccines production, etc. The complexity of these particle-fluid flow fields has, for many years, defied systematic investigations. A considerable change of attitude and progress have started in the last decade. Many fundamental aspects of centrifugal separation of mixtures have been formulated, analyzed, numerically simulated and experimentally verified [1]. The concomitant advance of supercomputing opens the possibility for implementing these achievements in large scale parametric studies and engineering-oriented simulations. The first part of the lecture is concerned with a short review of the methodology used in the analysis of separating suspensions and some typical results. However, as in any new field, every solution increases the appetite for additional knowledge on harder problems, and here an interesting aspect comes in: additional progress on problems of rotating suspensions seems to be hindered by the incomplete knowledge in the related classic theory of "single-phase" rotating fluids [2], in particular concerning the flow field generated by a moving particle. The second part of the lecture is concerned with this challenging issue.

2. FUNDAMENTAL MODELS AND RESULTS

The physico-mathematical framework used in most of the recent studies on suspensions is the continuum-mechanics approach. The numerous dispersed macroscopic particles lose their identities and are envisaged as a continuum— "the dispersed phase", which co-exists with another continuum—"the continuous phase", representing the embedding fluid. The "concentration" of the phases is usually represented by the volume fraction of the dispersed particles, α; evidently, the continuous phase occupies the volume fraction $1-\alpha$. We emphasize that even for quite dilute suspensions the interparticle distance, e^*, is of the order of magnitude of the particle size. In particular, for a suspension of spheres of radius a^*, simple volume considerations show that $e^* \approx 2.0a^*\alpha^{-\frac{1}{3}}$. Hereafter the asterisks denote dimensional variables. The notations are like in [1], and a short list is given at the end of this paper. The combination of the dispersed and continuous phases is regarded as the "mixture fluid". It is endeavored that the flow fields of these "phases" and "mixture" models reproduce the averaged motion of the physical system. To this end, the "averaged" equations are derived from basic principles. The resulting "two-fluid" model consists of two coupled sets of conservation equations. A more common variant, the "mixture" or "diffusion" model, uses one set of conservation equations for the whole mixture, supplemented by a "diffusion" equation to account for the internal changes of volume fraction (concentration) of the dispersed phase. Both models still require some constitutive closure assumptions: it is usually postulated that the generalized stresses are Newtonian with an "effective" viscosity, the interfacial "drift" force is given by a modified Stokesian drag formula, the pressures of the "phases" are related by the capillary law, etc.

The final set of conservation equations resembles a Navier-Stokes systems. Obviously, the hopelessly complicated original particle-fluid system has been cast in a tractable form by the continuum approach. Nevertheless, obtaining and verifying the solution of the averaged equations in a non-trivial configuration is still a formidable task. Details of solutions in typical gravity and centrifugal configurations were presented and discussed during the course. These included: settling in straight containers, settling in inclined containers (associated with the Boycott effect), centrifugal separation from initial solid-body rotation in axisymmetric straight and "inclined" containers, and the "spin-up" of a suspension from rest to (almost) solid-body rotation—but during which separation may occur— which has no counterpart in gravity settling. The details are long and therefore are not repeated here; the interested reader may find them in [1]. We emphasize that the abovementioned accepted postulates, concerning the interfacial "drag", $\mathbf{F}^*$, and the effective viscosity, $\mu^*_{\rm eff}$, are based actually on Stokes flow approximations, and have been tailored with gravity settling in mind. A key parameter in centrifugal separation is the (modified) particle Taylor number, $\beta = \frac{2}{9}\frac{\Omega^* a^{*2}}{\nu_0^*}$ which expresses the ratio of Coriolis to viscous forces on a particle. Evidently, in rapid centrifuges the Taylor number, β, of the dispersed particle is not small and the abovementioned accepted postulates concerning the interfacial "drag" and the effective viscosity (rheology) must be modified to account for the Coriolis effects that supplement and even dominate the assumed Stokes flow balances. Problems associated with these effects are still on the frontier of investigation, as explained below.

3. SOME FRONTIER TOPICS IN CENTRIFUGAL SEPARATION

In many current practical applications the value of the particle Taylor number, β, is small, because the dispersed particles are "small" and/or the rate of rotation is not "large". Nevertheless, the investigation of centrifuges with larger β may prove to be rewarding. (We may recall the study of supersonic flow at the time when airplanes attained about one tenth of the sonic speed.) It is convenient to use a cylindrical system r, θ, z co-rotating with the container (centrifuge) at the angular velocity $\Omega^* \hat{z}$. The velocity vector $\mathbf{v}$ has components u, v, w. For the quantitative analysis of the settling, it is essential to specify the relative velocity, $\mathbf{v}^*_R = \mathbf{v}^*_D - \mathbf{v}^*_C$. It can be argued that, as a first approximation, this results from a buoyancy-drift-Coriolis balance on each particle,

$$\rho^*_D \cdot 2\Omega^* \hat{z} \times \mathbf{v}_R = (\rho^*_D - \rho^*_C)\Omega^{*2} r^* \hat{r} + \mathbf{F}^*/(vol) \tag{1}$$

where (vol) is the volume of one particle and $\mathbf{F}^*$ is the hydrodynamic "drift" force on it due to the relative motion. For slow motion and $\beta \to 0$ we expect that the dominant behavior is close to that of a Stokesian flow, hence the hydrodynamic "drift" force is given by the conventional drag force result (for a sphere)

$$\mathbf{F}^* = -6\pi\nu^*_0 \rho^*_C a^* \mathbf{v}^*_R = -\frac{1}{\beta}\left(\frac{4}{3}\pi a^{*3}\rho^*_C\right)\Omega^* \mathbf{v}^*_R. \tag{2}$$

The drag-buoyancy balance then yields, for a spherical particle, the relative velocity correlation $\mathbf{v}^*_R \approx \varepsilon\beta\Omega^{*2} r^* \hat{r}$ in a centrifugal field, which is the counterpart of $\mathbf{v}^*_R = \varepsilon\frac{2}{9}(a^{*2}/\nu^*_0)g^*\hat{g}$ in the gravity field. Recall that $\varepsilon = (\rho^*_D - \rho^*_C)/\rho^*_C$. However, (1)-(2) also give a component of $\mathbf{v}^*_R$ in the $-\hat{\theta}$ direction, which has no counterpart in gravity settling. The modeling of centrifugal separation at not-very-small values of β is still a challenge. Actually, many of the most interesting, and perhaps useful, differences between the gravity and centrifugal separation appear for non-small β. This is because as β increases the influence of the Coriolis terms, which are not present in gravity settling, becomes dominant. We can list several gaps of knowledge in the theory of suspensions for non-small β, concerning:

(a) The internal (interfacial) "drag" force. Theoretical investigations [3]–[5] show that, for a spherical particle:

- For $\beta \ll 1$

$$\mathbf{F}^* = -6\pi\nu_0^*\rho_C^* a^* \{\mathbf{v}_R^* + \sqrt{\beta}[A(u_R^*\hat{\boldsymbol{r}} + v_R^*\hat{\boldsymbol{\theta}}) + Cw_R^*\hat{\boldsymbol{z}}] + \sqrt{\beta}B(-v_R^*\hat{\boldsymbol{r}} + u_R^*\hat{\boldsymbol{\theta}})\}, \quad (3)$$

where $A = 15/7\sqrt{2}, B = 9/5\sqrt{2}, C = 12/7\sqrt{2}$. The terms associated with the coefficients A and C predict an increase in the drag force due to the Coriolis effects. Moreover, a "lift" force, acting perpendicular to the direction of motion appears.

- For $\beta \to \infty$

$$\mathbf{F}^* = -\frac{8}{3}\pi^2 a^{*3}\rho_C^* \frac{1}{16+\pi^2}\Omega^*[(4u_R^* + \pi v_R^*)\hat{\boldsymbol{r}} + (4v_R^* - \pi u_R^*)\hat{\boldsymbol{\theta}}]. \quad (4)$$

We do not know what is the practical range of validity of these results, and how they are affected by the presence of other particles in the suspension, see below.

(b) The rheology, or the "effective viscosity". The well known correlations between viscosity and volume fraction (extensions of Einstein's formula) of the type $\mu_{\text{eff}}^*/\mu_0^* \approx \mu(\alpha) \approx (1+\frac{5}{2}\alpha)$, have been developed for, and tested in, non-rotating environments, $\beta = 0$. At large β Taylor column structures are bound to appear on the particles and it is not clear how this affects the internal stress distribution.

(c) The Ekman layer of the suspension. Near a boundary of the centrifuge which is not parallel to the axis of rotation, the internal flow is accommodated to the no-slip conditions in a region of thickness $\sim \delta_E^* = (\nu_0^*\Omega^*)^{1/2}$. We notice that $\sqrt{\beta} = (\sqrt{2}/3)a^*/\delta_E^*$. Thus, only if β is small, can the Ekman layer be considered as a region of "mixture".

3.1 THE INTERPRETATION OF CYLINDRICAL COMPARTMENT CENTRIFUGE EXPERIMENTS

Some practical dilemmas connected with the abovementioned gaps of knowledge show up in the analysis of the cylindrical compartment centrifuge ([6], [7]; [1], Sect. 7.5). The configuration is sketched in Fig. 1. The main flow is in the r, θ plane. If α in the suspension is a function of the time t^* only, then the buoyancy-drift-Coriolis balance, with the $\mathbf{F}^*$ from any of the previous formulas, gives the form:

$$\mathbf{v}_R^* = (\varepsilon\beta\Omega^* r^*)[U_R(t^*)\hat{\boldsymbol{r}} + \omega_R(t^*)\hat{\boldsymbol{\theta}}] \quad (5)$$

where U_R and ω_R depend on the specific behavior of the "drift" and of the correlation $\mu(\alpha)$. On the other hand, the continuity equation for the dispersed phase, when $\mathbf{v}_R^*$ is of the form (5) yields $\frac{d\alpha}{dt^*} = -2\varepsilon\beta\Omega^*\alpha(1-\alpha)U_R$, which is consistent with the assumption on α. The rate of settling (sedimentation) on the walls is also closely connected with $\mathbf{v}_R$.

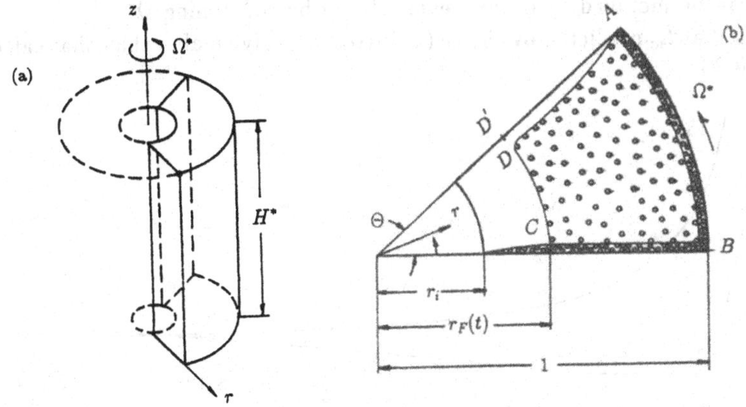

Figure 1: Cylindrical compartment centrifuge. (b) shows schematically the cross section during separation of a suspension of heavy particles, $\rho_D > \rho_C$; ADC is the interface between the pure fluid and the suspension domains.

Therefore, under some mild simplifying assumption, the time required for the complete separation --or settling - of the particles which were initially in the suspension can be calculated as

$$t^*_{sc} = \frac{1}{2U_R(1+\frac{1}{2\Theta}\frac{\omega_R}{U_R})} \ln\left[r_i^2 - \frac{1}{2\Theta}\frac{\omega_R}{U_R}(1-r_i^2)\right]\frac{1}{\varepsilon\beta\Omega^*} . \tag{6}$$

The final evaluation of (5) and (6) was performed with the assumption (2) in [6] and (3) in [1] Sect. 7.5. Some indicative graphs of the velocity components and the corresponding t_{sc}, vs. β, are displayed in Figs. 2 and 3.

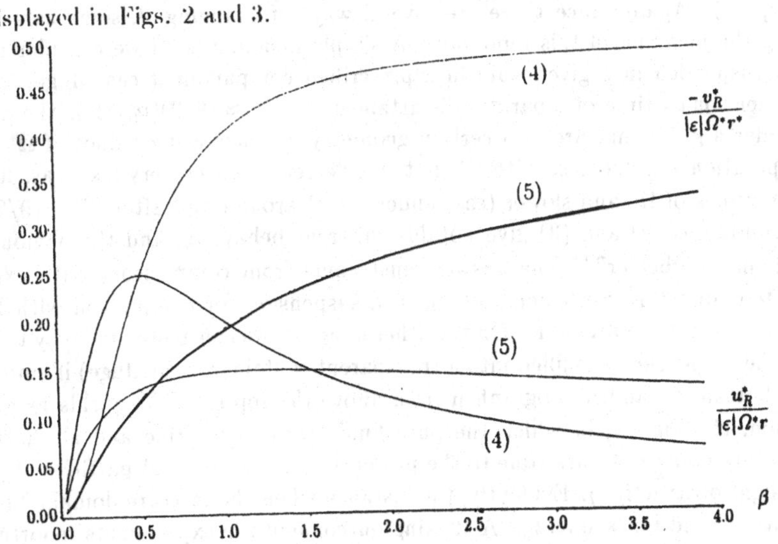

Figure 2: The scaled components of the relative velocity, see (8), vs. β, (in dilute limit where $\mu(\alpha) = 1$). The postulates (4) (Stokes) and (5) (Herron et al.) for the interfacial force were used.

Here the scaling highlights fixed ε and Ω^*, in which case the most obvious interpretation to various β is various particle sizes (e.g., we can change β from 0.25 to 1 by simply doubling a^*). However, β can be modified by other means, chiefly by controlling Ω^*.

Figure 3 indicates that t_{sc} predicted by choice (2) is considerably smaller than that calculated

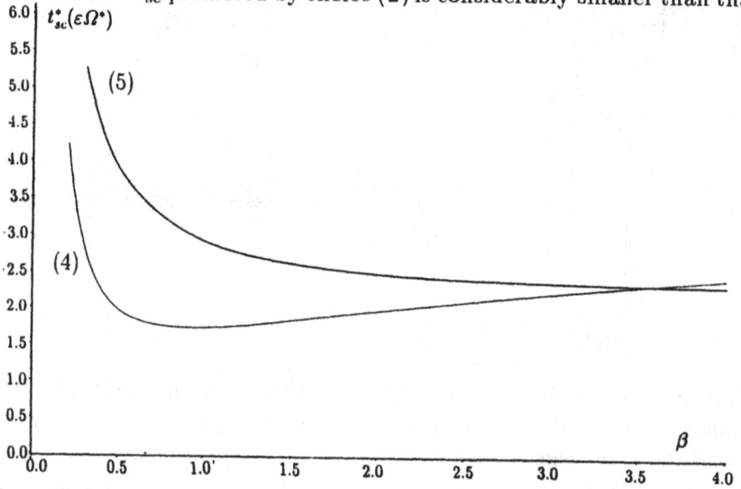

Figure 3: The scaled theoretical separation time in compartment centrifuge with $r_i = 0.2, \Theta = 30°$, see (10), vs. β, obtained with the postulates (4) (Stokes) and (5) (Herron et al.).

with (3) for the range $0.1 < \beta < 2$. An interesting quality of the present results is the minimum of $t^*_{sc}\Omega^*/|\varepsilon|$ predicted by (2) for $\beta \approx 1$. The physical interpretation follows from the variation of $\mathbf{v}_R$ with β, see Fig. 2: for $\beta \approx 1$, the azimuthal component of the settling velocity is already big while the radial component is still big, hence both the settling on the outer wall and on the side barrier are effective. No such minimum is obtained with the relationship (3). Again, since there are several ways of changing β, some care is required in grasping the meaning of this minimum. A simple meaning is: if we consider particles of various a^* suspended in a given fluid in a prescribed compartment centrifuge, the shortest actual (dimensional) time of separation is attained for $a^{*2} \approx (9/2)(\nu_0^*/\Omega^*)$. the other hand: if we consider a given mixture in a certain geometry, for increasing values of Ω^* the actual time of separation will decrease with Ω^*, but this decrease will be very fast (say, superlinear) for smaller values of Ω^* and slower (say, underlinear) around and after $\Omega^* = (9/2)(\nu_0^*/a^{*2})$. So the assumptions (2) and (3) give notably different behaviors, and the obvious question is: Which one is "better"? The answer must come from comparisons with experiments. Unfortunately, direct measurements of $\mathbf{v}_R$ in a suspension for comparison with Fig. 2 are very difficult and yet not available. On the other hand, t^*_{sc} was measured in [6] by the following method. The suspension was filled into a transparent container (centrifuge) illuminated from below the bottom lid and photographed from above the top lid. This yields pictures of the cross section, cf. Fig. 1, in which the pure fluid (water, glycerine and alcohol) is bright and the mature domain is dark (due to the presence of the suspended particles of polystyrol, which are light-obstructing). Evidently, the instance when the mixture domain "disappears" can be captured and marked as t^*_{sc}. A striking outcome of the experiments reported by [6] is the consistent detection of the abovementioned minimum in $t^*_{sc}\Omega^*/|\varepsilon|$.Moreover, the measured separation times for $0.05 \lesssim \beta \lesssim 5, \Theta = 30°, 45°, 60°$ were in good agreement with calculations based on (3) (with $\alpha(0) = 0.1$, $\mu(\alpha) = (1-\alpha)^{-2.7}$, see (3.)) and the observed interfaces were stable.

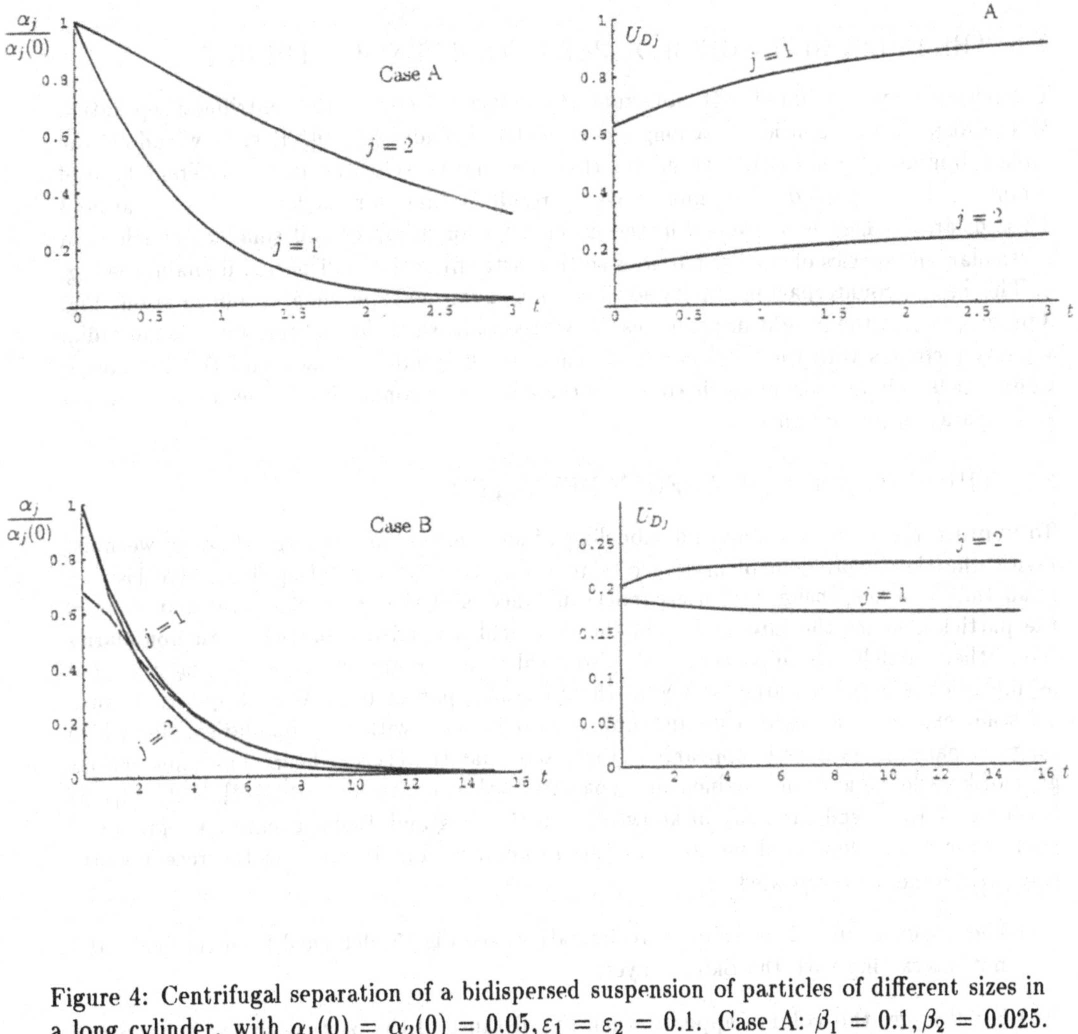

Figure 4: Centrifugal separation of a bidispersed suspension of particles of different sizes in a long cylinder, with $\alpha_1(0) = \alpha_2(0) = 0.05, \varepsilon_1 = \varepsilon_2 = 0.1$. Case A: $\beta_1 = 0.1, \beta_2 = 0.025$. Case B: $\beta_1 = 1.0, \beta_2 = 0.5$. The variation with time of the volume fraction and of the radial velocity for the components in the bidispersed region are shown. The doted line is the value in the adjacent left-behind monodispersed region (there is a "jump" via a kinematic shock). (Ungarish 1995). Time scaled with $(\varepsilon_1\beta_1\Omega^*)^{-1}$, velocity with $\varepsilon_1\beta_1\Omega^* r^*$.

The interesting conclusion is that the application of the simple "postulate" (2) for the relative velocity yielded quantitatively correct results for $\beta \lesssim 5$. This seems to be more than a coincidence, but the plausible explanation to the "validity" for non small β of (6), based on Stokesian drag, is still lacking. On the other hand, our attempt to incorporate the corrections of [3], (3), in the closure for $\mathbf{v}_R$ caused both qualitative and quantitative discrepancies with the reported experiment. Our observation is that at $\alpha \approx 0.1$ the interparticle distance is so small, about $0.9a^*$ according to (2.), that the appearance of the Taylor columns is suppressed. This may explain the loss of validity of (3), but cannot explain the validity of the Stokesian drag.

3.2 THE RELIABILITY OF BI-DISPERSED SUSPENSION THEORY

The peculiarities associated with non-small β are also reflected in the centrifugal separation of a bi-dispersed suspension in a long axisymmetric cylinder [8], [9]. Even if we adopt the Stokes drag assumption (2), it turns out that the relative velocities of the different type of components are coupled. The most striking result is that, for Taylor numbers β around 0.5 and larger, there is a reversal in the usual behavior of larger and smaller particles. In particular, the species of smaller particles settles faster from the bi-dispersed domain, see Fig. 4. This has no counterpart in gravity settling, and, so far, no experimental confirmation. Fig. 3 provides again the insight into this result: we see that, with the assumption (2), the radial velocity increases with the particle size a^* when $\beta < 0.5$, but decreases with this parameter when $\beta > 0.5$. In an axisymmetric container there is a close connection between radial motion and separation performance.

3.3 THE SINGLE PHASE THEORY REVISITED

To improve the understanding and modeling of suspensions in the large β range we must revisit the classic problem of motion of single particle(s) in a rotating fluid. We keep in mind that in a suspension the interparticle distance is of the order of several diameters of the particles, hence the interactions of the flow field around one particle with boundaries (i.e., other particles) is important. We also recall that we are interested in practical (not asymptotically small or large) values of the governing parameters. We ask questions such as: what exactly is a Taylor column? how does it interact with the boundaries? for which range of parameters does it appear? It turns out that the classic theory has some serious gaps of knowledge and unexplained discrepancies with experiments ([10], [11], [12], [13]). It is necessary to extend the body of knowledge on the drag and Taylor column. Progress was made concerning slow axial motion. In this respect, we briefly mention the recent works [14]–[17]. Some novel knowledge:

(a) The Taylor column is a region of recirculation, see Fig. 5, detached from the body and not interacting with the Ekman layer;

(b) For β large, this column appears when the distance of the particle to the boundary is larger than $0.4\beta a^*$, and is fully developed for $1.1\beta a^*$;

(c) If influence from boundaries is neglected, a Taylor column appears when $\beta > 8.2$. (From previous investigation, it is known that the length of this column is $\sim 0.23\beta a^*$.)

(d) The drag calculations by the "linear theory" (i.e., with convection of momentum neglected, or zero Rossby number) are applicable to practical ranges of parameters. The discrepancy with experiments in the "short cylinder" case, usually attributed to non-linear terms, seems to be rather a viscous effect.

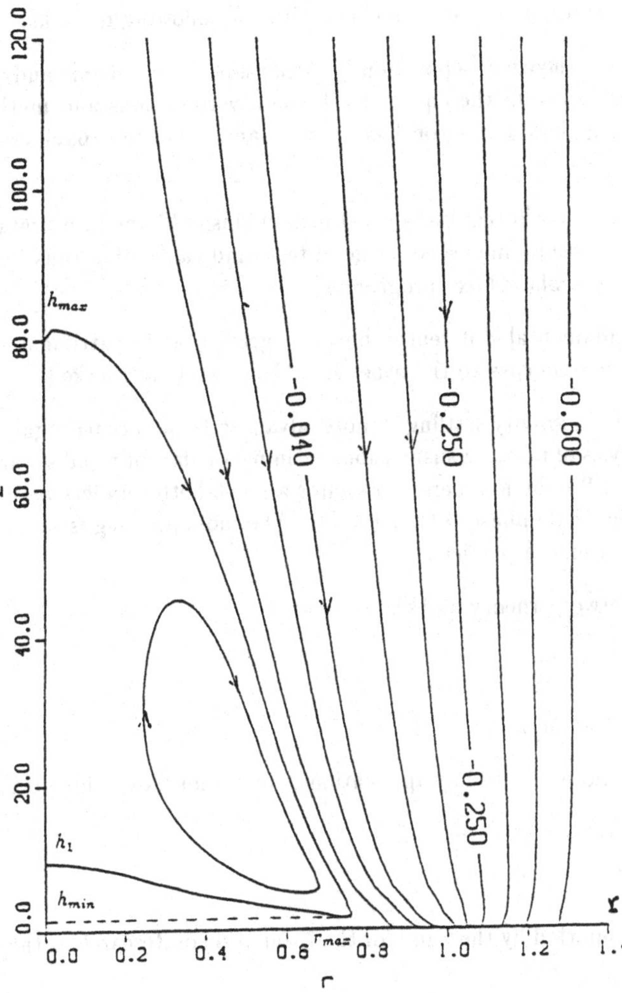

Figure 5: The forward Taylor column on a particle moving axially. Streamlines as seen from the center of a disk particle; a symmetric behavior occurs in the rear side. (Vedensky & Ungarish 1994). Lengths are scaled with radius of particle. $\beta = 356$.

These new results still await experimental corroborations. Unfortunately, concerning the lateral motion, little progress was made. Here we mention the paper [18] about a disk moving in its own plane. Since there is no drag on such a body, the relevance of these results to separation problems is quite restricted. The analysis of lateral motion of a particle of finite thickness (preferably, a sphere or ellipsoid), which throws additional light on the results (3)-(4), remains an important open challenge.

4. CONCLUDING REMARKS

In view of the discussed material, we may conclude with the following remarks.

- The foundations of the theory of separation of suspensions seem reliable and productive. We have gained confidence in the equations, boundary conditions and methods of approach; we acquired understanding on basic results and seen the capability of treating complex flow fields.

- Application of the theory still requires a great deal of "insight" and interpretation of the results. The available results may serve as good tests and calibration runs for computer codes, in particular for "black-box" programs.

- There are some fundamental differences between gravity and centrifugal separation. "Extrapolations" from one case to the other should be carefully checked.

- The "state of the art" in gravity settling is more advanced than in centrifugal separation. Progress in the analysis of rotating suspensions — in particular for rapid separation with non-small Taylor and Rossby numbers -- requires a much better understanding of some "classic" problems in single-phase rotating fluids. The most pressing issues concern the flow-field and force around a particle:
  - discrepancies between theory and experiments,
  - stability,
  - lateral motion,
  - Taylor columns interaction.

- It makes sense to promote this field by applications, simulations, experiments and further research.

ACKNOWLEDGMENT

This work was partially supported by the Fund for the Promotion of Research at the Technion.

REFERENCES

1. Ungarish, M.: Hydrodynamics of Suspensions: Fundamentals of Centrifugal and Gravity Separation. Springer-Verlag, Berlin 1993.

2. Greenspan, H.P.: The theory of rotating fluids. Cambridge U. Press, Cambridge 1968.

3. Herron, I.S., Davis, H. & Bretherton, F.P.: On sedimentation of a sphere in a centrifuge, J. Fluid Mech., 62 (1975), 209-234.

4. Childress, S.: The slow motion of a sphere in a rotating viscous fluid, J. Fluid. Mech., 20 (1964), 305-314.

5. Stewartson, K.: On the slow motion of an ellipsoid in a rotating fluid, Quart, J. Mech. Appl. Math. 6 (1953), 141-162.

6. Schaflinger, U., Köppl, A. & Filipczak, G.: Sedimentation in cylindrical centrifuges with compartments, Ing. Arch., 56 (1986), 321–331.

7. Dahlkild, A. A. & Greenspan, H. P.: On the flow of a rotating mixture in a sectioned cylinder, J. Fluid Mech., 198 (1989), 155–175.

8. Ungarish, M. & Greenspan, H. P.: On centrifugal separation of particles of two different sizes, Int. J. Multiphase Flow, 10 (1984), 133-148.

9. Ungarish, M.: On the modeling and investigation of polydispersed rotating suspensions, Int. J. Multiphase Flow, 21 (1995), 262-284.

10. Moore, D.W. & Saffman, P.G.: The structure of free vertical shear layers in a rotating fluid and the motion produced by a slowly rising body. Trans. Roy. Soc. Lond. A 264 (1969), 597-634.

11. Maxworthy, T.: The observed motion of a sphere through a short, rotating cylinder of fluid, J. Fluid Mech. 31 (1968), 643-655.

12. Maxworthy, T.: The flow created by a sphere moving along the axis of a rotating, slightly-viscous fluid, J. Fluid Mech. 40 (1970), 453-479.

13. Karanfilian, S. K. & Kotas, T. J.: Motion of a spherical particle in a liquid rotating as a solid body, Proc. R. Soc. Lond. A, 376 (1981), 525-544.

14. Vedensky, D & Ungarish, M.: The motion generated by a slowly rising disk in an unbounded rotating fluid for arbitrary Taylor number, J. Fluid Mech., 262 (1994), 1-26.

15. Tanzosh, J. P. & Stone, H. A.: Motion of a rigid particle in a rotating viscous flow: An integral equation approach, J. Fluid. Mech., 275 (1994), 225-256.

16. Ungarish, M. & Vedensky, D.: The Motion of a Rising Disk in a Rotating Axially Bounded Fluid for Large Taylor Number, J. Fluid Mech., 291 (1995), 1-32.

17. Ungarish, M.: Some shear-layer and inertial modifications of the geostrophic drag on a slowly rising particle or drop in a rotating fluid, J. Fluid Mech., to appear (1996).

18. Tanzosh, J. P. & Stone, H. A.: Transverse motion of a disk through a rotating viscous fluid, J. Fluid. Mech., 301 (1995), 295-324.

NOTATION LIST

a^* - radius of dispersed particle
$\mathbf{F}^*$ - hydrodynamic force on dispersed particle
g^* - gravity acceleration
r, θ, z - cylindrical coordinates
u, v, w - velocity components in cylindrical coordinates
$\mathbf{v}^*$ - mass velocity vector; $\mathbf{v}_R = \mathbf{v}_D - \mathbf{v}_C$
α - volume fraction of dispersed phase; $\alpha(0)$ - initial value
β - Taylor number of particle (Coriolis/viscous forces)
$\varepsilon = (\rho_D - \rho_C)/\rho_C$
$\mu(\alpha)$ - effective viscosity ratio function
ν_0^* - kinematic viscosity of suspending fluid
ρ^* - density
Ω^* - absolute angular velocity of system (centrifuge)
Subscripts : D, C - dispersed, continuous phase

Upper asterisk denotes dimensional variable

Printed in the United States
By Bookmasters